Lean Management System LMS:2012

A Framework for Continual Lean Improvement

Lean Management System LMS:2012

A Framework for Continual Lean Improvement

William A. Levinson

CRC Press is an imprint of the
Taylor & Francis Group, an **informa** business
A PRODUCTIVITY PRESS BOOK

CRC Press
Taylor & Francis Group
6000 Broken Sound Parkway NW, Suite 300
Boca Raton, FL 33487-2742

Printed in the United States of America on acid-free paper
Version Date: 20120518

International Standard Book Number: 978-1-4665-0537-7 (Paperback)

Library of Congress Cataloging-in-Publication Data

Levinson, William A., 1957-
Lean management system LMS:2012 : a framework for continual lean improvement / William A. Levinson.
p. cm.
Includes bibliographical references and index.
ISBN 978-1-4665-0537-7 (pbk.)
1. Lean manufacturing. 2. Continuous improvement process. 3. Costs, Industrial. I. Title.

TS155.L3673 2012
658.5'33--dc23 2012017891

Visit the Taylor & Francis Web site at
http://www.taylorandfrancis.com

and the CRC Press Web site at
http://www.crcpress.com

Contents

Preface

Why LMS:2012?

This publication's purpose is to meet a long-unfilled need for a *Lean management system* (LMS) with which to standardize and organize Lean manufacturing, Lean enterprise, and Lean service activities and programs. Its key elements are

1. A simple yet comprehensive set of Lean key performance indicators that can identify and measure all conceivable forms of waste (muda).
2. A proactive continuous improvement cycle that seeks out waste instead of reacting to waste that comes to its attention.
3. An unofficial (and therefore customizable) standard around which organizations can structure a comprehensive and synergistic set of proactive activities to remove all forms of waste from their operations and supply chain, and to ensure systematic continuity of future improvement activities. The unofficial standard is available by itself from the publisher at http://www.crcpress.com/product/isbn/9781466505377. (Disclaimer: No part of this unofficial standard is affiliated with or authorized by the International Standards Organization (ISO) and its series on management standards.)

Most Lean manufacturing books focus on operations research and industrial engineering concepts like kaizen events, workplace rearrangement to improve efficiencies, visual controls, pull production systems, heijunka or production leveling, and so on, whose primary purpose is to reduce cycle time and inventory. LMS:2012 adds material and energy considerations and therefore implements the key elements of the ISO 14001 environmental management system (EMS) and ISO 50001 energy management system (EnMS) standards. The recommended *material and energy balance* analytical technique gives material and energy wastes absolutely no place to hide and should therefore meet and may even exceed these standards' basic requirements.

The Ford Motor Company meanwhile exemplified the merits of ISO 14001 and ISO 50001 in the language of money 70 or more years before these standards' creation. No environmental protection laws prevented Henry Ford from throwing into the River Rouge whatever waste he couldn't send up his smokestacks. Norwood's (1931, 124) *Ford: Men and Methods*, however, has an entire chapter called "Converting Waste into Millions," which translates into "tens of millions" in today's money. Distillation of waste wood alone yielded $12,000 a day in wood chemicals and charcoal, which (assuming that the distillation plant did not work on weekends) added up to $3 million a year in the money of 1926.

Achievement of similar results today requires treatment of ISO 14001 and ISO 50001 as servants and not masters. If the organization's goal is to comply with a standard to "get the certificate," the standard is the master and also a costly and time-consuming annoyance. These standards become valuable servants when the organization treats them as frameworks to identify and eliminate material and energy waste, respectively. This book's goal is to equip the reader to do exactly that.

LMS:2012 is emphatically not a substitute for any official standard; however, if there are no wastes of material, there are no ISO 14001 environmental aspects. If there is no waste of energy, there can be no gap between the organization's energy performance and any energy targets it might have set under ISO 50001. The practical achievability of this ideal or perfect state is not the issue, although zero material waste and therefore zero pollutant emissions may be close to achievable as proven by Ford more than 80 years ago. Subaru (2011, website) adds, "In 2004, our Subaru of Indiana manufacturing plant became the first automotive assembly facility to be 'zero landfill,' meaning all waste is recycled or turned into electricity." The issue is whether the organization has procedures and techniques to identify and address all material and energy wastes. If it does, and this publication offers a comprehensive set, it should be able to answer most audit questions that relate to ISO 14001 and ISO 50001.

Why does the manufacturing and even the service sector need a standard, whether official or unofficial, for LMSs? The first chapter cites contemporary authorities on this issue, and the Ford Motor Company's unprecedented initial success followed by its seemingly inexplicable decline provides the ultimate and final answer.

OVERVIEW

I

Chapter 1

The Need for a Lean Management Standard

The bottom-line results of Lean manufacturing were incontrovertible almost 100 years ago, and long before the leaders at Toyota even heard of it. Charles Buxton Going's preface to Arnold and Faurote (1915) describes these results in the language of money—the same language that gains buy-in from executives, CEOs, and other decision makers.

> Ford's success has startled the country, almost the world, financially, industrially, mechanically. It exhibits in higher degree than most persons would have thought possible the seemingly contradictory requirements of true efficiency, which are: constant increase of quality, great increase of pay to the workers, repeated reduction in cost to the consumer. And with these appears, as at once cause and effect, an absolutely incredible enlargement of output reaching something like one hundred fold in less than ten years, and an enormous profit to the manufacturer.

There is nothing about "enormous profit to the manufacturer" with which management can possibly disagree, and "great increase of pay to the workers" should be equally self-explanatory to hourly workers and, if there is one, the union. This book will later address the need for management and labor buy-in in detail and provide information with which to achieve it.

Scientific progress should build on scientific progress, so what became of this manufacturing Golden Age and industrial Garden of Eden? Why do

American families now struggle on two incomes when a single job at the Ford Motor Company could once support a middle-class family comfortably? Hogg (2011, 104) provides an answer:

> However, after 20 years, less than 10 percent of the companies that have initiated lean are exploiting it fully. A breakthrough is needed to explode this percentage and reach the hundreds of thousands of North American organizations that are now only using the tools and ignoring the huge gains that system-wide leverage can bring. The current uncertain turmoil may be preparing us for the launch of a universal lean breakthrough.

System-wide leverage comes from a Lean manufacturing standard, or at least something that can be used as one until an official standard comes along. The "universal Lean breakthrough" that Hogg describes is this book's explicit mission, and its basic elements were available 90 years ago:

> I am now most interested in fully demonstrating that the ideas we have put into practice are capable of the largest application—that they have nothing peculiarly to do with motor cars or tractors but form something in the nature of a universal code. I am quite certain that it is the natural code and I want to demonstrate it so thoroughly that it will be accepted, not as a new idea, but as a natural code (Ford and Crowther, 1922, 3).

Smith (2011b, 11) reinforces this concept with the statement, "Lean is not so much about what you see but is more about how you think." The way to reproduce Ford's success is therefore not to become preoccupied with a particular Lean tool. When the only tool one has is a hammer, every problem starts to look like a nail. People and organizations must therefore understand and internalize Ford's entire universal code, which synergizes the laws of science (including a wide array of Lean tools) with the laws of human and organizational behavior.

This section has already shown this universal code's achievement of "an absolutely incredible enlargement of output reaching something like 100-fold in less than 10 years, and an enormous profit to the manufacturer" with regard to automobiles, and it also delivered world-class results in mining, rail transportation, and health care. The bottom-line results are incontrovertible; the methods and principles are very easy to understand, and they can serve as the foundation of a universal Lean breakthrough that will restore economic prosperity and affluence to the United States. *This book's explicit mission is to deploy Ford's universal code, along with subsequent developments by Toyota and other Lean practitioners, in the format of an unofficial standard.*

If the off-the-shelf answer has been available for 90 years, though, why is it not in widespread use throughout the country? Hogg (2011) already showed that use of Lean tools is not equivalent to use of a Lean system, and Blanchard (2007) supports this assessment. A 2007 *Industry Week* (Tadsen, 2008) survey, which allowed multiple responses, reported that 69.6 percent of U.S. manufacturers used Lean manufacturing. Another 34.2 percent used Total Quality Management (TQM), 29.0 percent Six Sigma, and 17.0 percent the Toyota Production System. 14.4 percent used the Theory of Constraints, a pull production system that is admittedly only one aspect of a comprehensive Lean system. Only 11.6 percent used none of the listed methods, which means 88.4 percent used one or more. When it came to results, however, only 26.1 percent—fewer than a third of those that used recognized quality and productivity improvement methods—had made significant progress toward world-class status or had fully achieved it. Tadsen (2008, Webinar) elaborates:

> Use of lean tools is not the same as being a lean organization. The goal is a continuous improvement culture. There is a "glass ceiling" between mere use of tools and techniques, and a continuous improvement culture. It is culture that assures that a company that is on top today will be there 20 years from now.

The title of the presentation was incidentally "Planning and Executing a Sustainable Lean Transformation," and the word *sustainable* deserves emphasis. The Ford Motor Company had a very conspicuous Lean culture through at least the early 1930s, which it then lost and had to relearn decades later. This shows that even culture is not enough, and that only a Lean management *system* can ensure that a company that is on top one day (as Ford was in the 1920s) will remain there permanently.

Weeks (2011, 24) adds a similar observation about the need to systemize or even standardize Six Sigma: "Many Six Sigma implementations were not integrated into the goals and objectives of companies, and Six Sigma simply became one of the goals and objectives." ISO saw a need for a standard and published ISO 13053:2011, "Quantitative methods in process improvement—Six Sigma." Per *Quality Digest* (2011), this standard includes the following:

- **Part 1: DMAIC methodology** describes the five-phased methodology—define, measure, analyze, improve, control (DMAIC)—and recommends best practices, including on the roles, expertise, and training of personnel involved in such projects.
- **Part 2: Tools and techniques** describes tools and techniques, illustrated by fact sheets, to be used at each phase of the DMAIC approach.

The subtitle adds explicitly, "DMAIC in ISO format should reduce fragmentation and harmonize best practices." ISO has not, however, published or even, as far as we know, drafted an explicit standard for comprehensive Lean management systems. LMS:2012 is therefore designed to meet this need unofficially, which leaves the user free to modify it to meet industry-specific or company-specific needs. The next section will provide the strongest argument for a Lean management standard.

The Serpent in the Industrial Garden of Eden

The Ford Motor Company developed what most people now call the Toyota Production System (Levinson, 2002a), along with many other leading-edge management, quality, safety, and productivity techniques. The results described by Charles Going's Foreword to Arnold and Faurote (1915, iii) were due to clearly identifiable Lean manufacturing practices at the Highland Park plant *along with organizational behavior considerations that gained buy-in and engagement from the workforce* (emphasis added). Ford had by the 1920s a deeply ingrained Lean enterprise culture that included supply chain management and empowerment of workers to identify and eliminate waste on sight. Workers at the River Rouge plant could even stop the line if there were trouble—a practice that Toyota made famous much later. Ford's Lean system manifested itself through the following results:

- Ford quadrupled his workers' wages in 20 years.
- Ford bought out his minority shareholders for $12,500 for each $100 16 years after he founded his company.

- The Ford Motor Company was directly responsible for making the United States the wealthiest and most powerful nation on earth. It transformed the country from a primarily agrarian society to an industrial society while giving us the 40-hour five-day workweek along with a prosperous middle class.

Ford demanded standardization of every job in his industry, including mining, railroads, and so on, but he did not standardize the amazing management *system* that served him and his company so well. Ford, like Alexander the Great long before him, never planned for a time when his empire would have to function without him—or simply grew too large for a single person to oversee effectively. The company's decline was therefore directly traceable to the loss of key management personnel during the early 1940s. Ford suffered a series of strokes that rendered him unable to lead his company, his son Edsel (a capable leader despite what was later named for him) died, and production manager Charles Sorensen retired. The effect was immediately foreseeable:

> Sorensen had been the plaster that held the [River Rouge] plant together. When word came through that he was leaving, real panic swept through the Rouge. When Sorensen left, the Rouge lost its soul. (Bennett, 1951, 298)

The company then had to get Henry Ford II, a naval officer with almost no experience with his grandfather's business system, to assume a nominal leadership role. The actual business direction then came from exactly the kind of people whom Henry Ford said should never be allowed to run any business because their eye was always on the dollar and rarely on the job that earned it.

> In the decade following World War II, Ford's Whiz Kids created a corporate culture based on a financial paradigm, in which virtually every business decision was a function only of profitability. (Hoyer, 2001, 35)

The situation was so pathetic by 1982 that when a group of Ford executives visited Japan to learn from the Japanese what their own company had taught the Japanese, they did not even recognize their own company's methods. "One Japanese executive referred repeatedly to 'the book.' When Ford executives asked about the book, he responded: 'It's Henry Ford's book of course—your company's book'" (Stuelpnagel, 1993, 91). Shirouzu (2001) later reported how James Padilla, the group vice president in charge of global manufacturing, visited Mazda to learn "Japanese" Lean manufacturing methods. Upon Padilla's return, Shirouzu wrote:

> Here, and throughout Ford's global operations, Mr. Padilla is trying to entice assembly workers and engineers to abandon nearly all they know about the mass manufacturing system that Henry Ford brought to life about 90 years ago. (A1, A6)

The techniques Mr. Padilla wanted his people to learn were, however, really the inventions or at least developments of Henry Ford and his coworkers, but neither Padilla nor anybody else at Ford apparently recognized them as such! The reference continues:

> It [worker empowerment] isn't a brand-new idea in American industry. But it is one that some workers said they regard suspiciously because they worry that Mr. Padilla ultimately intends to increase efficiency and eliminate jobs. (A1, A6)

The original Ford culture encouraged workers to identify all forms of waste and to not worry about going through channels to eliminate the waste. The culture of 2001 had therefore backslid to a Luddite mentality in which any improvement was perceived as a threat to jobs instead of a path to higher wages—a devastating self-destructive paradigm that this book will address in detail. Shirouzu (2001) adds that a worker objected to an initiative to change the workplace layout to eliminate the need to take about 2,000 steps per shift, again because the perception was that workers would be discharged if efficiency improved. The company's founder wrote explicitly that pedestrianism is not a paying line of work because it produces no value, and that no job should require anybody to take more than one step in any direction. "Save ten steps a day for each of twelve thousand employees and you will have saved fifty miles of wasted motion and misspent energy" (Ford and Crowther, 1922, 77).

If that is what the company's founder said about 10 steps per employee per day, what should the workforce have said about 2,000? This comes to somewhat more than a mile, and it takes a person about 20 minutes, or 1/24 of an eight-hour shift, to walk this far at a normal pace. The worker's pay is therefore about 4 percent less than what it would be without this waste motion. Management can gain buy-in from the workforce by showing how higher efficiencies lead to higher wages and often less physical effort as well. Chapter 2 will elaborate on this in detail.

Sinclair (1937, 81) shows, however, that the Luddites were of the company's own creation:

> Twenty men who had been making a certain part would see a new machine brought in and set up [at the River Rouge plant], and one of them would be taught to operate it and do the work of the 20. The other 19 wouldn't be fired right away—there appeared to be a rule against that. The foreman would put them at other work, and presently he would start to "ride" them, and the men would know exactly what that meant.

This statement is enormously revealing as to why workers who would have never dreamed of unionizing when Ford was actually managing the company demanded a union in the late 1930s. LMS:2012 includes a no-layoff policy as a mandatory provision of a Lean management system, along with a "no restrictive work rules" policy for labor. Productivity improvements followed by layoffs equals Luddites, and Ford therefore apparently implemented a rule against discharging anybody when productivity improved. The people to whom he left the management of his company during the 1930s found ways to circumvent the no-layoff rule and therefore forfeited the loyalty of the workforce.

Excessive reliance on an individual or core group of people to hold a corporate culture together is meanwhile far from unique to Ford. Per Glader (2006, B1),

> Often, an operation becomes Lean because of a single personality—a top executive who sets an example and is relentless about eliminating waste. But when the personality leaves, does the Lean mentality survive in the next generation? "Hewlett Packard had the H-P way. The founders are gone, and the H-P way is mostly gone," says Jeffrey Liker, a management expert and professor of engineering at the University of Michigan, Ann Arbor.
>
> Nucor's personality was Ken Iverson…

Ford and Crowther (1922, 85–86) meanwhile reveal a glaring flaw in Ford's own system that might have easily contributed to the system's collapse in the early 1940s. ISO 9001 implementers can meanwhile use this excerpt to educate a workforce on the reason for that standard's extensive documentation requirements, and Chapter 10 discusses in detail the need for documentation. The bottom line is that, if it isn't written down, it didn't happen.

> The factory keeps no record of experiments. The foremen and superintendents remember what has been done. If a certain method has formerly been tried and failed, somebody will remember it—but I am not particularly anxious for the men to remember what someone else has tried to do in the past, for then we might quickly accumulate far too many things that could not be done. That is one of the troubles with extensive records. If you keep on recording all of your failures you will shortly have a list showing that there is nothing left for you to try—whereas it by no means follows because one man has failed in a certain method that another man will not succeed.

Standard scientific research practice is of course to record everything whether successful or not. Records of failures can prevent others from wasting resources by repeating them and can also lead to successes by encouraging future researchers to try alternatives. In addition, had Ford kept written records of experiments, they would have served as valuable case studies in his trade school and would be useful for training workers and engineers today. The case studies Ford cites in his books are enormously instructive and provide more than a hint of the enormously useful knowledge that has been lost through his company's failure to record its experiments. Ford was, however, correct that people who "know" that something cannot be done are right, at least as far as they are concerned personally.

We can therefore conclude that, if Ford had developed an LMS standard to ensure continuity of his Lean enterprise methods, his company would still be on top of the world and the Toyota production system would be rightly known as the Ford production system. The purpose of LMS:2012 is to make the Lean management system independent of key personnel, just as ISO 9001:2008 makes the quality system independent of key personnel, and to ensure continuity of the programs and activities.

The next section will show why Six Sigma is not up to the necessary job, although its elements are certainly compatible with Lean manufacturing.

Lean, Six Sigma, or Both?

Why not simply register to ISO 13053:2011 instead of implementing a comprehensive LMS? First, a comprehensive Lean system can do everything Six Sigma can plus a lot more. Levinson (2011a) points out that Six Sigma was apparently not up to the task of keeping jobs at three of its foremost implementers—General Electric (GE), Motorola, and Maytag—in the United States. Six Sigma, therefore, cannot offer the excuse that companies that move manufacturing jobs offshore are simply not using it properly.

Maytag, for example, developed and trademarked a Lean Six Sigma program during the 1990s. Miller (2002, website) reported however, that "Maytag Corp. plans to close a 1,600-worker refrigerator factory in rural Galesburg, Illinois, transferring much of its production to Mexico, where labor is cheaper." Long (2003) adds that Motorola closed its Harvard, Illinois, plant about six years after it began operations. Many if not most GE consumer appliances are made offshore.

This is not to say that these companies would have been better off without Six Sigma, because it contains many proven quality and productivity improvement tools. The shipment of jobs offshore for cheap wages is nonetheless an open admission, at least by the standards of Ford and his contemporaries, that a productivity improvement program simply does not have what it takes to deliver consistent world-class results. Henry Towne, past president of the American Society of Manufacturing Engineers, stated more than 100 years ago that what we now call Lean manufacturing was developed explicitly to keep jobs in the United States:

> We are justly proud of the high wage rates which prevail throughout our country, and jealous of any interference with them by the products of the cheaper labor of other countries. To maintain this condition, to strengthen our control of home markets, and, above all, to broaden our opportunities in foreign markets where we must compete with the products of other industrial nations, we should welcome and encourage every influence tending to increase the efficiency of our productive processes. (Foreword to Taylor, 1911b)

To this, Ford and Crowther (1926, 116) added that the decision as to where to build a new plant was a function of access to transportation and cheap energy—two criteria from which access to low-wage labor is conspicuously absent. A quality and productivity improvement program or system that is not up to the job of keeping jobs in the United States is therefore inadequate on its face.

This does not mean that Six Sigma cannot and does not deliver impressive results when used correctly. The problem appears to be that many forms of waste or muda are simply invisible to Six Sigma's traditional metrics, and that Six Sigma is not sufficiently comprehensive to deliver world-class results. It is a matter of historical record that Ford's system did, and to it we can add new sciences (including, for example, industrial statistics) that developed subsequently.

Second, there is not a mutually exclusive choice between TQM, Lean manufacturing, and Six Sigma. The latter two rely on the existence of the former because it is very difficult, to say the least, to do either Lean or Six Sigma when gages are out of calibration, products and product data are not traceable, and so on. Most Six Sigma techniques are meanwhile TQM techniques as well. Levinson (2011a) contends that the label "Six Sigma" does with TQM methods what a legendary "soup stone" does with traditional stew ingredients in the story about stone soup—it gets people to use them. *Consistent and systematic use of adequate productivity improvement methods will achieve far more than inconsistent and fragmented application of the best methods on earth.*

Third, the Automotive Industry Action Group's continuous quality improvement (CQI-10) (2006) applies to a much broader range of problems than Six Sigma's DMAIC. CQI-10 is similar to the Ford Motor Company's Team-Oriented Problem Solving, Eight Disciplines (TOPS-8D), which has the same advantages over DMAIC. Workers with high school educations find it easy to understand, and it does not focus on quantitative performance metrics to which many obvious problems and improvement opportunities are invisible. CQI-10 and 8D can, however, incorporate Six Sigma's most advanced quantitative methods whenever they are applicable.

LMS:2012 encourages and systemizes the application of Lean, quality, Six Sigma, and anything else that delivers the necessary results. A focus on Lean manufacturing or Lean enterprise is therefore synergistic with Six Sigma, and nothing stops an organization that pursues this recommended course from meeting the requirements of ISO 13053 as well. The next section will address the thought process behind the development of LMS:2012. It is first necessary to distinguish between critical to quality (CTQ) and critical to Lean (CTL) characteristics.

Critical to Quality versus Critical to Lean

An LMS complements and is synergistic with a quality management system (QMS), but there is also an enormous difference between them. The QMS focuses on *critical to quality* (CTQ) characteristics, which are almost if not entirely uniformly intrinsic to the product or service. In the case of physical products, these include feature dimensions, roughness, durability, hardness, purity, reliability, and other characteristics that a supplier or customer can measure on any individual unit. Services are less tangible, but their CTQ characteristics are also measurable; they include customer satisfaction, on-time delivery, and so on. ISO 9001:2008 and ISO/TS 16949 focus on the processes that deliver the product or service, but their emphasis is almost entirely on CTQ characteristics.

This book introduces a new term, *critical to Lean* (CTL) characteristics, and contends that there are a total of four for any product or service. Chapter 2 will go into them in extensive detail, but Table 1.1 summarizes them here.

CTQ characteristics are almost universally measurable on an individual product or service activity, but *CTL characteristics are inherent to the process that delivers the product or service.* Consider two pieces of work that are equally perfect in terms of meeting specifications; that is, their CTQ characteristics are identical. If one comes from a pull production system with almost zero inventory and the other from an inventory-laden system, the first will be less expensive and better able to meet customer delivery schedules. If one comes from a process that recycles what are normally considered consumable materials, while the other comes from a process that throws them away after a single use, the first will again be less expensive

Table 1.1 Critical to Lean Characteristics (and Lean Key Performance Indicators)

1. Waste of the time of things: specifically the product or service (cycle time), or an asset *for which paying work is available*
2. Waste of the time of people
3. Waste of materials → Environmental Management System and ISO 14001
4. Waste of energy → Energy Management System and ISO 50001

and therefore more competitive. This book will later cite Frank Gilbreth's improvement of a bricklaying process, in which the end quality of two walls might be identical in terms of CTQ characteristics but one is far less expensive because of the more efficient process behind it. This underscores the need for the Lean system to focus on the CTL characteristics.

CTL considerations may, however, support CTQ considerations. Inventory gives defects a place to hide, so its elimination (reduction of the waste of the time of things) also helps prevent poor quality. The purpose of Design for Manufacture (DFM) is to make the product easier and simpler to make, which often reduces its cost and increases its quality simultaneously. The QMSs and LMSs should take advantage of such synergies as much as possible.

The next section will address specific considerations in the development of LMS:2012.

LMS:2012 Development Considerations

LMS:2012 has been developed with the following considerations in mind:

1. Implementation of LMS:2012 should also implement all key elements of ISO 14001 and ISO 50001, noting that Lean manufacturing seeks to eliminate wastes of materials and energy, respectively.
   a. As shown by Fraser and Kutky (2011) and Block and Marash (1999) (discussed later in this section), a company that meets the requirements of ISO 9001 is most of the way to ISO 14001. LMS:2012 similarly relies on the presence of a QMS, and it also addresses very explicitly wastes of materials and energy.
   b. LMS:2012 goes beyond the requirements of ISO 14001, which calls for identification of environmental aspects, or considerations that can affect the environment. It calls for identification of *all* material wastes, including those (such as recyclable scrap metal) that do not endanger the environment or incur costs for pollution control and environmental protection. This more than encompasses everything that ISO 14001 would call an environmental aspect.
2. LMS:2012 should be synergistic and compatible with widely used QMS standards like ISO 9001:2008 and ISO/TS 16949:2002. Duplication of effort, such as the preparation of separate sets of documents with the opportunity for conflicting directives, is absolutely to be avoided.

a. Excessive complexity, such as mapping to Malcolm Baldrige and Six Sigma (neither of which focuses primarily on Lean), is to be avoided, although some of their provisions can be referenced where appropriate. As an example, Six Sigma's Supplier, Input, Process, Output, Customer (SIPOC) concept supports examination of entire processes and value streams as opposed to individual functions. The Society of Automotive Engineers' (1999) SAE J4001 also provides some useful ideas.
b. Internal audits of the LMS should piggyback onto audits of the other standards. LMS:2012 will therefore be organized exactly like ISO 9001:2008. As an example, the LMS requirement for closed loop *proactive* action will piggyback onto the ISO requirements for closed loop preventive and corrective action.
c. Fraser and Kutky (2011) state that a company that is registered to ISO 9001 is 60 percent of the way to registration to ISO 14001 and/or OSHAS 18001. This reinforces the desirability of structuring LMS:2012 like ISO 9001 wherever possible and to defer to ISO 9001 (or ISO/TS 16949) for all common requirements like document control, record retention, gage calibration, and so on.
d. Block and Marash (1999) add explicitly that ISO 14001 can be integrated into a QMS and that a single manual can cover the requirements of both systems. The reference states, "The use of existing quality management procedures to fulfill ISO 14001 requirements eliminates redundancy and ensures consistency" (p. 5). The same concept applies to the LMS.

3. LMS:2012 should, like all official manufacturing-related ISO and AIAG standards, be the organization's servant and not its master. Organizations that register to ISO 9001:2008 to "get the certificate because our customers want us to have it" and regard their QMSs as costly and time-consuming annoyances get exactly that: costly and time-consuming annoyances. Those that use ISO 9001 to ensure quality along with continuity of improvement efforts find themselves with a valuable system that frees the workday from genuine costly and time-consuming annoyances like scrap, rework, and customer returns.
4. LMS:2012 should promote narrative audits as opposed to yes/no responses, checklist responses, and Likert scale numerical ratings. Rudyard Kipling's "I Keep Six Honest Serving-Men" (1897) is highly instructive and very applicable to all audit-related activity.

> I keep six honest serving-men
> (They taught me all I knew);
> Their names are What and Why and When
> and How and Where and Who.

5. Planning and assessment questions should include at least one of these words if possible. A question like "How does management promote a Lean culture?" evokes a more detailed response than "Does management promote a Lean culture?" "What are the synergies between the LMS and QMS?" requires a more detailed response, and evokes more thoughts and ideas for improvement, than "Are the LMS and QMS synergistic?" "Who" in the context of quality (and by implication LMS) audits relates almost universally to a position instead of a named individual.
6. This consideration supports the principle that a system should serve the organization instead of the organization serving the system. The idea is not to check off enough "yes" responses or get a sufficiently high numerical rating to pass an audit and get a certificate. The idea is to use the system to drive continuous improvement. A question like "How does management promote a Lean culture?" and supporting subquestions like "How do performance metrics encourage people to eliminate waste? What dysfunctional metrics encourage wasteful purchasing practices or production of unnecessary inventory?" encourage people to think about ways to promote a Lean culture.

The next section discusses the organization of LMS:2012 into provisions, assessment questions, and supplemental information.

Organization and Implementation of LMS:2012

LMS:2012 is presented as an unofficial or voluntary standard whose structure allows it to piggyback onto ISO 9001:2008, but it is vital for the organization to know not only what it is doing (following the standard) but why it is doing it. The purpose of the standard is to systemize and deploy the relatively simple and easily understandable foundation of a comprehensive LMS. This consists of two basic elements: (1) a handful of simple key performance indicators (KPIs) and (2) a proactive continuous improvement cycle that applies these KPIs to all operating processes, including service activities.

Proactive means the improvement cycle sets out to look for trouble in the form of waste or muda instead of standing ready to react to trouble.

LMS:2012 Section I: Foundation of a Comprehensive Lean Management System

Chapter 2 will lay the groundwork by defining four recommended Lean KPIs whose purpose is to identify and measure *every possible form of waste* in *any manufacturing or service operation*. These are wastes of (1) time of things (products or services, as in cycle time), (2) time of people, (3) materials, and (4) energy. The chapter will show how these KPIs encompass the Toyota Production System's Seven Wastes and also the basic metrics of Goldratt's Theory of Constraints.

Carbon emissions are not among the recommended KPIs, and Chapter 2 cites extensive reasons for not using them. These reasons can and should be shared with supply chain partners and the public as a whole. Any action taken to reduce waste of energy will automatically reduce carbon emissions if the energy comes from fossil fuels. This book's position is, however, that extraordinary efforts to reduce carbon emissions for their own sake or, even worse, the purchase of carbon offsets are not a responsible use of stakeholder and supply chain resources.

The combination of the recommended KPIs with Chapter 3's proactive continuous improvement cycle, Isolate, Measure, Assess, Improve, Standardize (IMAIS) systemizes the identification of all waste for subsequent elimination. The four recommended KPIs and IMAIS are designed explicitly to give an organization everything it needs to gain the maximum possible benefit not only from a Lean enterprise system but also from ISO 14001 and ISO 50001.

LMS:2012 Section II: Voluntary and Customizable Lean Management System Standard

The second section contains LMS:2012 itself: a structure and framework against which internal auditors can assess an LMS. Each numbered section follows the organization of ISO 9001:2008 to allow piggybacking, and each is designed to deploy the recommended KPIs (or their equivalents) to management responsibility, resource management, product or service realization, measurement and analysis, and environmental and energy considerations.

1. The provisions of LMS:2012 itself are directly below the numbered heading. The words *shall* and *must* refer to mandatory provisions against which the LMS is to be audited. *Shall* and *must* will be italicized to emphasize mandatory provisions.
   a. The words *can* and *may* refer to optional or recommended features of the LMS. As an example, "The LMS can and should be incorporated into an existing quality management system (QMS) to avoid duplication of effort and conflicting directives." It doesn't have to, but maintaining a separate LMS is likely to lead to unnecessary work.
   b. In many cases the provision will say only, "LMS:2012 defers to the prevailing QMS standard" or "LMS documents and records *shall* be treated exactly like QMS documents and records." Lean manufacturing cannot, for example, work without calibrated gages and instruments, but calibration is already covered by ISO 9001:2008 and ISO/TS 16949. There is of course no reason to have one document control and record retention system for the QMS and another for the LMS.
2. The assessment question lists are not formal provisions of LMS:2012. Their purpose is to help the auditors determine whether the provision is being fulfilled and also to serve as guides in developing and improving the LMS.
   a. The organization has the ultimate responsibility for determining how it wants to meet the provisions. It may therefore add or remove questions to guide the development and implementation of the LMS that best suits its needs.
3. Acronyms for provisions that support ISO 14001, ISO 50001, and OSHAS 18001 include:
   a. LMS/E = Energy Management System
   b. LMS/M = Materials (Environmental) Management System
   c. LMS/S = Safety and Industrial Hygiene

Figure 1.1 shows the relationship between the Lean KPIs, the IMAIS improvement cycle, and the unofficial standard.

LMS:2012 Section III: Supplementary Detail

The third section then goes into extensive detail for each LMS provision including (1) the reason for doing it and (2) methods, techniques, and

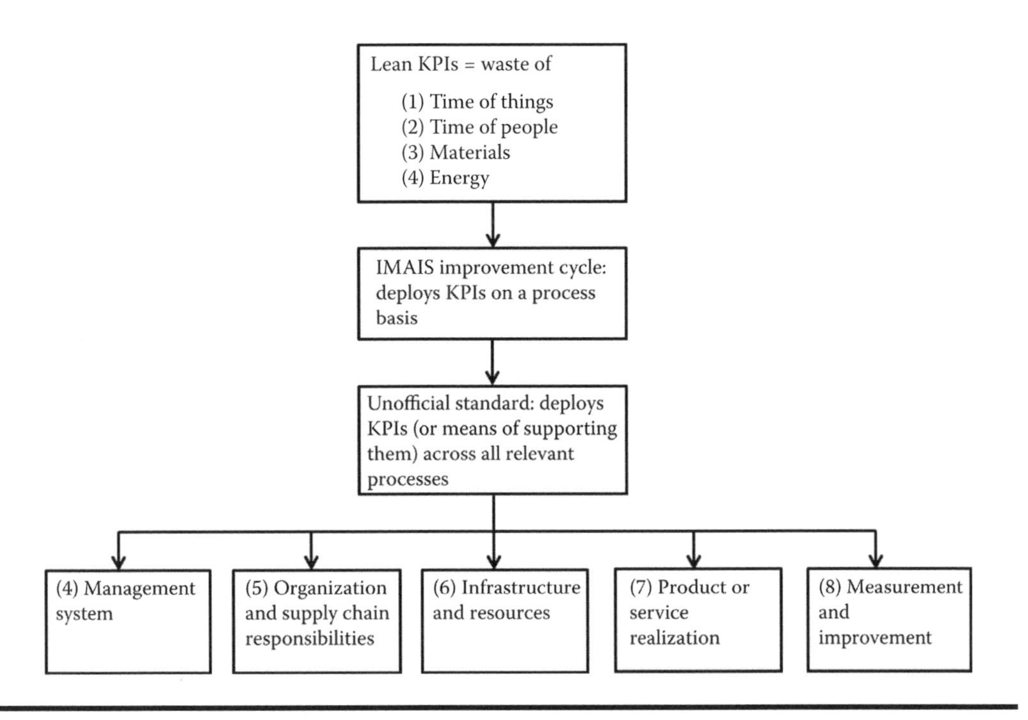

Figure 1.1 Lean KPIs, IMAIS, and unofficial standard.

thought processes for implementation. The examples may or may not be applicable directly to the user's industry, but the thought process is important in all cases.

Chapter 2

Lean Key Performance Indicators

Frederick the Great's advice that one who tries to protect everything protects nothing applies to performance indicators as well. An organization that tries to measure everything similarly ends up measuring nothing because a workforce that has to pay attention to a huge laundry list of key performance indicators (KPIs) cannot devote to each one the attention it might or might not deserve. The eventual result is that most or all soon become meaningless. "Five or less" is a frequently quoted rule of thumb.

Table 1.1 listed four critical to Lean (CTL) characteristics: (1) waste of the time of things, (2) waste of the time of people, (3) waste of materials, and (4) waste of energy. A comprehensive Lean system must therefore measure these as KPIs in some manner. Organizations that do not already have these KPIs in place are well advised to adopt them because they were central to the Ford Motor Company's phenomenal success. Ford and Crowther (1922, 329) recognized that there are exactly three kinds of waste: waste of time, waste of materials, and waste of energy.

> You can waste time, you can waste labor, you can waste material—and that is about all. ...Time, energy, and material are worth more than money, because they cannot be purchased by money. Not one hour of yesterday, nor one hour of today can be bought back. Not one ounce of energy can be bought back. Material wasted, is wasted beyond recovery.

It is, however, useful to break down "waste of time" into two categories: waste of the time of things (cycle time, plus assets for which paying work is available) and waste of the time of people (labor). The result is four KPIs that encompass the Toyota Production System's Seven Wastes, the recommended metrics of Eliyahu Goldratt's Theory of Constraints, and also the role of almost every Lean technique and procedure like Just-in-Time (JIT) and single-minute exchange of die (SMED). These four KPIs can expose *all* wastes in any manufacturing or even service supply chain, including wastes that are not visible to Six Sigma's traditional metrics.

Every form of waste in an *operating process* is measurable with one or more of these KPIs. Idle plant, land, and/or equipment tie up capital but, as they do not participate in any process, they cannot create any of the indicated wastes and are therefore invisible to these KPIs. This is something to keep in mind for official standards like ISO 9001 as well because an idle asset that is not part of a process is not subject to audit. The focus of LMS:2012 is, however, on operating processes, where the preponderance of the waste is likely to be found.

Waste of the Time of Things

This includes all time in which no tool or equipment performs value-adding work on the product, or when an asset *for which value-adding work* is available is idle or underutilized. If for example a passenger airplane's maintenance, boarding, and deplaning times can be reduced sufficiently, an extra flight can be added every day to increase the asset's revenue. On the other hand, no asset should produce unusable inventory just to keep it busy; hence the qualifier, "for which value-adding work is available." Goldratt's Theory of Constraints says this condition applies only to the constraint or capacity-constraining resource in a factory situation.

A tool adds value only when it transforms the product. Time spent in holding and releasing the work does not count as value-adding. Neither does setup, which is the reason for SMED.

Far greater wastes of time occur when the product is waiting on a shelf or is in transit on a truck or container ship. This kind of waste is very likely to occur at a process interface, which returns attention to the issue of handoffs between processes or supply chain partners. The principle is that the work should be in continuous motion toward a tool if it is not actually being processed by one, with the motion in question being as short and continuous as possible. A chemical factory in which liquids and gases either are inside process units (such as reactors, distillation columns, absorption towers, and so on) where value is added or flow in a pipe from one unit to another is the ideal model. The counterpart of a pipe in a factory for discrete units is a conveyor or work slide, on which the work is in continuous motion. Every employee should question the status of every piece of work that is not either (1) being acted upon by a tool or (2) in motion on a conveyor belt, work slide, or less ideally a forklift or hand truck.

Quality practitioners recognize variation in critical to quality (CTQ) characteristics as a source of nonconformances, but *variation in processing and material transfer times* is a major reason for the waste of the time of things. Ford and Crowther (1922, 143) recognized explicitly the effect of this variation on inventory and therefore cycle time.

> If transportation were perfect and an even flow of materials could be assured, it would not be necessary to carry any stock whatsoever. The carloads of raw materials would arrive on schedule and in the planned order and amounts, and go from the railway cars into production. That would save a great deal of money, for it would give a very rapid turnover and thus decrease the amount of money tied up in materials. With bad transportation one has to carry larger stocks.

Hopp and Spearman (2000, 270) present the Kingman Equation (Equation (2.1)), which puts Ford's statement into mathematical terms. It shows how he could have indeed run a balanced factory at close to 100 percent capacity without the consequences illustrated by Goldratt's (1992, 103–112) matchsticks and dice exercise. Henry Ford (Ford and Crowther, 1922, 82) claims explicitly to have done this: "The idea is that a man must not be hurried in his work—he must have every second necessary but not a single unnecessary second."

$$CT_q = \left(\frac{c_a^2 + c_e^2}{2}\right)\left(\frac{u}{1-u}\right)t_e \qquad (2.1)$$

where:

CT_q = cycle time in queue
c_a = coefficient of variation (standard deviation divided by mean) for arrivals at the operation
c_e = coefficient of variation in effective processing time
u = utilization
t_e = effective processing time

If "an even flow of materials could be assured" then the coefficient of variation for arrivals is zero. If there is no variation in effective processing times, the indicated term for that also is zero. This makes the cycle time in queue, and hence the inventory, zero for any utilization of less than 100 percent: "It would not be necessary to carry any stock whatsoever."

If, however, there is any variation whatsoever, cycle time and hence inventory increase enormously with utilization. In Goldratt's exercise, matchsticks (or anything else) that represent products are processed by a line of people who roll a single die for each unit of time. The person can pass to the next operation the number of items indicated by the die *provided that they are available.* Since time lost at a constraint can never be made up, temporary shortages caused by the variation become permanent production losses. High die rolls that would presumably offset low ones are wasted when the inventory at a workstation is less than the die roll. Goldratt describes this condition as *starvation of the constraint,* i.e., the constraint goes idle for lack of work. The effect is that some workstations gather enormous piles of inventory while others are practically empty.

Ford, however, removed processing variation from the work through subdivision of labor to the extent that one person placed a nut on a bolt but did not tighten it. Performance of both tasks would have required the worker to pick up and put down a wrench each time, which is both non-value-adding and a source of extra variation in processing time. Work slides, conveyors, and even work cells eliminated most variation inherent in material transfer time. The KPI "waste of the time of things" focuses attention on all these considerations.

The magnitude of this kind of waste cannot be overemphasized. Shingo (2009, 224–230) reports that conversion of a lot-based engine bearing manufacturing process to single-unit flow reduced the work in process inventory by 99.5 percent. If the throughput remained constant, Little's Law (cycle time = inventory divided by throughput) shows that the cycle time also must have been reduced 99.5 percent. The ability to carry far less inventory, and to deliver products to order instead of to forecast—and short cycle times are a prerequisite for delivery to order—is a decisive competitive advantage. Cycle time accounting (Levinson, 2007, 80–82) is a valuable technique for identifying waste of the time of things.

Waste of the Time of Things: Cycle Time Accounting

AIAG (2006, 192–193) describes *workflow analysis*, whose purpose is to identify value-adding and non-value-adding activities. It is synergistic with basic process flowcharting, which is one of the seven traditional quality improvement tools. (The others are the tally or check sheet, histogram, Pareto chart, scatter diagram, cause-and-effect diagram, and control chart.) It then assigns each activity to one of the following six categories, of which only *operation* adds value whereas *inspection* and *decision* may assist or enable value:

1. Operation
2. Transportation
3. Inspection
4. Delay or temporary storage
5. Storage
6. Decision

As with any other metrics or performance indicators, it is desirable to reduce the categories to the minimum necessary to classify the activities. Shingo (1987, 162) cites Frank Gilbreth's set of only four: processing, inspection, transportation, and delays. *Processing* must, however, encompass both value-adding transformation of the product and non-value-adding (or at best value-enabling) activities such as handling the work, adjusting the tool, and so on. Taylor (1911b, 171) breaks down a lathe operation (value-adding under both the AIAG and Gilbreth criteria) into non-value-adding "operations connected with preparing to machine work on lathes and with removing work to floor after it has been machined" and value-adding "operations connected with machining work on lathes." The former include handling and setup activities with prefixes like "putting in" and "taking out" whereas the latter describe activities in which the tool is in contact with the work.

Levinson (2007, 78–80) concludes that a minimum of five categories is required for workflow analysis. It is difficult to envision any piece of work that is not in one of these five states at any particular time.

1. Transformation
2. Handling
3. Transportation
4. Inspection
5. Delay

Transformation is the only activity that adds value, whereas handling and inspection may assist or enable value. Transportation is at best a necessary evil, whereas delay is clearly non-value-adding. Figure 2.1 shows that a Gantt chart can display how work in process spends time (perhaps an average as determined from numerous jobs) at a batch heat treatment operation at Ford's Highland Park plant as depicted by Arnold and Faurote (1915, 85–86). Times other than those stated by the reference—4 hours for annealing and 6.5 hours for removal and cooling—are assumed.

> The blank is stiff from the finishing rolls of the plate-steel mill, and is placed at once on a steel-roller gravity-incline and carried to the annealing ovens, 1500 degrees F, where the blanks are piled, one hundred and fifty on each of the six oven cars, which fill the oven with nine hundred blanks. As at present worked, the six cars go into the annealing oven at 8:00 p.m., remain until 12:00 midnight, are then withdrawn and left in piles on the cars until 6:30 a.m., where they are cool enough to be worked in the drawing press.

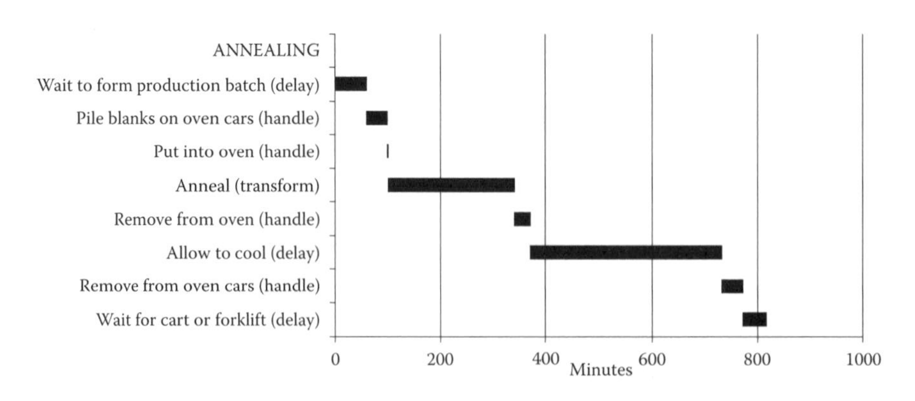

Figure 2.1 Cycle time accounting, Gantt chart format: batch annealing.

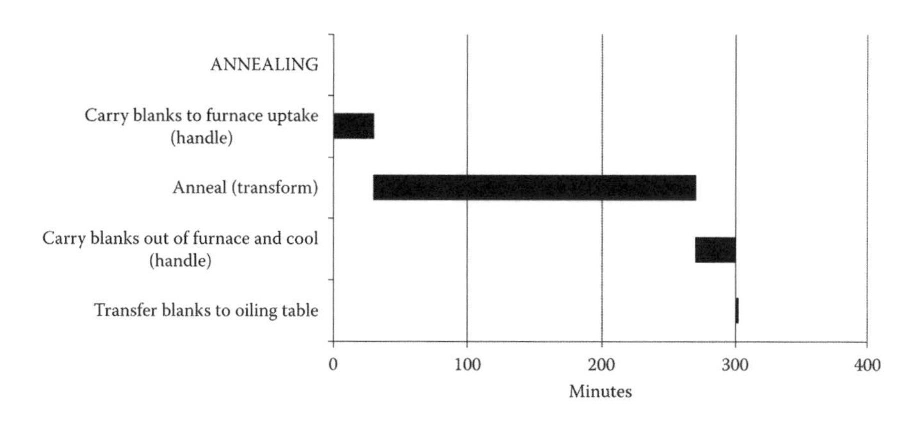

Figure 2.2 Cycle time accounting: continuous flow annealing.

The idea of color-coding the bars, with delays in red (no value whatsoever) and handling and transportation in yellow (necessary evils), comes to mind immediately. This format compels all non-value-adding time to become highly visible. In this case, "wait to form production batch" and "wait for cart or forklift" would attract immediate attention, whereas workers might also wonder why it takes something like six hours for the metal blanks to cool. The fact that they are "left in piles" on the oven cars, where they interfere with each other's heat transfer to the environment, provides a likely answer.

Figure 2.2 illustrates the same process after conversion to continuous flow (Arnold and Faurote, 1915, 86–87). It is assumed that the annealing process still requires four hours, but the figure shows the elimination of substantial non-value-adding cycle time.

> This first annealing practice will soon be obsolete. A furnace now under construction is served by an endless chain moving up and down, which is fitted with pendulum blank-carriers to take the blanks individually as they come through the press die, carry them upward about 60 feet in the furnace uptake, giving ample heating time, and then carry the blanks downward 60 feet in the open air, giving plenty of cooling time before the blanks reach the oiling table.

Also, remember that process interfaces and handoffs are frequent sources of waste as well as quality problems, and that the time necessary to move the work to and from the annealing operation is non-value-adding cycle time. Cycle time accounting is meanwhile compatible with the spaghetti diagram, which accounts for transportation inside a factory.

Work cells, in which tools and equipment are in the sequence of operations instead of individual departments, eliminate the transportation time and possibly some of the handling as well. The Ford Motor Company recognized this issue almost 100 years ago, as shown by Arnold and Faurote (1915, 38–39):

> Most of all, however, the Ford engineers have taxed the convolutions of their brain surfaces to shorten the lines of natural work-travel on the factory floors, first by crowding machine-tools together far closer than I have elsewhere seen machine-tools placed, and next by first finding the shortest possible lines of production travel of every car component, integral or assembled, and then placing every production agent needed either directly in that shortest line, or as near that line as possible, to the extent of placing even the brazing fires where most travel-saving advantage demands.

Transformation processes cannot, however, be taken for granted as value-adding. Schonberger (1986, 83) describes how Omark eliminated a heat treatment operation through selection of a steel alloy that did not require it. There are also "transformations" that are, upon closer inspection, 100 percent rework of problems that were built into previous operations. Ford and Crowther (1926, 67–68) described a straightening process for axles that did not cool uniformly after heat treatment, which sounds like a process or operation but is actually 100 percent rework. A centrifugal hardening machine was introduced that allowed the shafts to cool evenly and avoid entirely the need for the straightening operation. Deburring also is a form of rework, even though people normally take it for granted as a transformation process. If practical material removal processes that do not generate burrs are available, deburring becomes unnecessary.

In summary, cycle-time accounting requires classification of all time the work spends in the process (or series of processes) to identify time in which no value is added to the work. Reduction or elimination of this non-value-adding time enables shorter cycle times and all the advantages that come with them, such as less inventory and the ability to make to order instead of forecast.

The often enormous waste of the time of things cannot be overemphasized. Standard and Davis (1999, 61) use the analogy of a golf game in which value-adding work takes place only when the club head (tool) is in contact with the ball (part). This is about 1.8 seconds out of a typical 90-stroke 4-hour game. *The proportion of value-adding time to non-value-adding time is often little better in many manufacturing processes.*

Imai (1997, 22–23) explains the same concept as follows: "There is far too much muda between the value-adding moments. We should seek to realize a series of processes in which we can concentrate on each value-adding process—Bang! Bang! Bang!—and eliminate intervening downtime." In many cases, the value-adding moment is a literal "bang" from a press or stamping tool. The basic idea is that value is added only when the tool is doing something to the part while everything else (including loading, unloading, and orienting the part) is non-value-adding.

This chapter will show later that Frank Gilbreth, the father of motion efficiency, got many of his ideas from watching military drills in which the product was a literal value-adding "Bang!" Armies of Gilbreth's era used breech-loading rifles, but soldiers had within living memory used muzzle-loading weapons in which every action but the volley itself was setup or handling. Drills and equipment were both designed to minimize this setup as much as possible. The gunpowder came in a paper-wrapped cartridge that eliminated the need for the soldier to measure out each powder charge, i.e., it externalized this part of the setup process. The loading process began when the soldier bit open the cartridge, poured some of the powder into the priming pan, and dumped the rest down the barrel with the bullet. The cartridge (package) itself formed the wadding, which not only improved the motion efficiency of the entire process, but also used the packaging instead of throwing it away.

Drillmasters meanwhile made every possible effort to eliminate from the loading drill any motion or action that did not contribute to shorter cycle times and therefore more rapid volleys. The similarity of von Steuben's (1779) "Regulations for the Order and Discipline of the Troops of the United States" to the training within industry (TWI) *job breakdown sheet* is remarkable, as shown by Table 2.1. Von Steuben's original begins with a loaded musket, but the example assumes the weapon has just been fired.

Von Steuben did not include the reasons or explanations, presumably because men who were familiar with muskets already knew them. Table 2.1 adds sample reasons *and also classifies each action into one of the categories that cycle time accounting uses.* Note also that von Steuben includes the number of motions required for each step, an assessment that suggests clearly the modern concept of motion efficiency.

This exercise emphasizes Imai's contention that the value-adding "Bang!" may indeed make up only a tiny fraction of any process' total cycle time. The number of motions required to handle the rammer and the musket itself made the development of a breech-loading weapon the Holy Grail of small-arms development. The effort necessary to ram a bullet down a rifled

Table 2.1 Musket Loading Drill as Job Breakdown Sheet

Step (What to do)	*Important Points (How to do it)*	*Reason (Why do it?)*
Half-cock Firelock! 1 motion.	Half bend the cock briskly, bringing down the elbow to the butt of the firelock.	Handle (tool) Half cock exposes the pan for loading but the musket will not fire if the trigger is pulled accidentally.
Handle, Cartridge! 1 motion.	Bring your right hand short round to your pouch, slapping it hard, seize the cartridge, and bring it with a quick motion to your mouth, bite the top off down to the powder, covering it instantly with your thumb, and bring the hand as low as the chin, with the elbow down.	Handle (product) Covering the cartridge prevents accidental spillage of the powder and also protects it from moisture if present. This makes misfires less likely.
Prime ! 1 motion.	Shake the powder into the pan, and covering the cartridge again, place the three last fingers behind the hammer, with the elbow up.	Handle (product)
Shut, Pan! 2 motions.	1st. Shut your pan briskly, bringing down the elbow to the butt of the firelock, holding the cartridge fast in your hand. 2d. Turn the piece nimbly round before you to the loading position, with the lock to the front, and the muzzle at the height of the chin, bringing the right hand up under the muzzle; both feet being kept fast in this motion.	Handle (tool) Placement of the lock to the front (away from the soldier) presents the ramrod, which is opposite the lock, to the soldier.
Charge with Cartridge! 2 motions.	1st. Turn up your hand and put the cartridge into the muzzle, shaking the powder into the barrel. 2d. Turning the stock a little towards you, place your right hand closed, with a quick and strong motion, upon the butt of the rammer, the thumb upwards, and the elbow down.	Handle (product) The cartridge itself forms the wadding.

Table 2.1 Musket Loading Drill as Job Breakdown Sheet (continued)

Step (What to do)	*Important Points (How to do it)*	*Reason (Why do it?)*
Draw, Rammer! 2 motions.	1st. Draw your rammer with a quick motion half out, seizing it instantly at the muzzle back-handed. 2d. Draw it quite out, turn it, and enter it into the muzzle.	Handle (tool)
Ram down, Cartridge! 1 motion.	Ram the cartridge well down the barrel, and instantly recovering and seizing the rammer back-handed by the middle, draw it quite out, turn it, and enter it as far as the lower pipe, placing at the same time the edge of the hand on the butt-end of the rammer.	Handle (product) This combines the ramming operation with withdrawal of the rammer into a single motion.
Return, Rammer! 1 motion.	Thrust the rammer home, and instantly bring up the piece with the left hand to the shoulder, seizing it at the same time with the right hand under the cock, keeping the left hand at the swell, and turning the body square to the front.	Handle (tool)
Poise, Firelock! 2 motions.	1st. With the left hand turn the firelock briskly, bringing the lock to the front. At the same instant, seize it with the right hand just below the lock, keeping the piece perpendicular. 2d. With a quick motion bring up the firelock from the shoulder directly before the face, and seize it with the left hand just above the lock, so that the little finger may rest upon the feather spring, and the thumb lie on the stock; the left hand must be of an equal height with the eyes.	Handle (tool)

(continued)

Table 2.1 Musket Loading Drill as Job Breakdown Sheet (continued)

Step (What to do)	*Important Points (How to do it)*	*Reason (Why do it?)*
Cock Firelock! 2 motions.	1st. Turn the barrel opposite your face, and place your thumb upon the cock, raising the elbow square at this motion. 2d. Cock the firelock by drawing down the elbow, immediately placing your thumb upon the breech-pin, and the fingers under the guard.	Handle (tool)
Take Aim! 1 motion.	Step back about six inches with the right foot, bringing the left toe to the front; at the same time drop the muzzle, and bring up the butt-end of the firelock against your right shoulder; place the left hand forward on the swell of the stock, and the fore-finger of the right hand before the trigger; sinking the muzzle a little below a level, and with the right eye looking along the barrel.	Handle (tool)
Fire! 1 motion.	Pull the trigger briskly, and immediately after bringing up the right foot, come to the priming position, placing heels even, with the right toe pointing to the right, the lock opposite the right breast, the muzzle directly to the front and as high as the hat, the left hand just forward of the feather-spring, holding the piece firm and steady; and at the same time seize the cock with the fore-finger and thumb of the right hand, the back of the hand turned up.	Transform (product) The last action puts the hand and musket in the correct position for "Half-cock Firelock!"

barrel meanwhile divided foot soldiers into line infantry whose smoothbore muskets could fire four or more shots per minute (albeit very inaccurately beyond 70 or 80 paces) and riflemen who could fire only two shots per minute, but to far greater distances. The flintlock musket itself was, however, an improvement on the matchlocks of the sixteenth and early seventeenth centuries, which required additional handling operations to remove the burning cord of slow match before priming the pan (to avoid having the powder flask explode in the musketeer's hand) and replace it to touch off the charge.

This section has already shown that cycle time accounting seeks to categorize all the time the job spends in the factory or supply chain, with the production control system or (less ideally) manual observation by somebody with a stopwatch as sources of data. Every action or non-action must be timed, and these include the following:

- Value-adding time during which a tool is transforming the product.
- Waiting for an operator.
- Waiting for a tool.
- Waiting to form a transfer batch.
- Waiting for transportation (forklift, hand truck, or whatever).
- Transportation.
- Waiting for more parts to fill batch equipment such as an oven, furnace, or chemical bath.
- Waiting to get out of a batch at a single-unit work station. This consideration and the previous one show why single-unit processing, which approximates the continuous flow of a chemical factory, is best.

The job breakdown sheet in Table 2.1 along with the classification of each action as transformation, handling, inspection, transportation, or delay suggests its extension to the general subject of task efficiencies and the elimination of waste of the time of people as well as things. Division of labor also reduces variation in processing time, which (per Goldratt's Theory of Constraints [TOC]) increases inventory and cycle time while it reduces throughput.

Division of Labor and Variation Reduction

Ford and Crowther (1922, 83) wrote, "The man who places a part does not fasten it—the part may not be fully in place until after several operations later. The man who puts in a bolt does not tighten it." This sounds on its face like Ford wanted to reduce workers to the equivalent of automata, each of whom was entrusted with only the simplest task. It actually allowed Ford to do what the matchsticks and dice exercise in Goldratt's *The Goal* showed to be impossible: run a balanced factory at close to 100 percent capacity. This is easy to understand from Equation (2.2).

Equation (2.2): Total time and variance for a series of n tasks

$$T = \sum_{i=1}^{n} t_i \text{ and } \sigma^2 = \sum_{i=1}^{n} \sigma_i^2. \tag{2.2}$$

Table 2.2 applies this to the placement and tightening of a bolt as performed by one worker.

Subdivision of the job as described by Ford eliminates the time *and also the variance* of the non-value-adding steps "pick up wrench" and "put down wrench." Arnold and Faurote (1915, 106–110) describe how subdivision of a piston and rod assembly operation doubled its productivity and, with the addition of an inspector, improved quality:

> This piston-assembling job teaches two lessons of first importance. The first is that there are great savings in labor to be made by splitting operations to the extent that the workman does not need to change the position of his feet, and the second lesson is that a work-slide so located that the workman can drop his completed operation out of his hand in a certain place, without any search for a place of deposit, and also can reach to a certain place and there find his next job under his hand, is also a very important time-saver.

Table 2.2 Placement and Tightening of a Bolt

Step	*Classification*
Place bolt	Handle
Pick up wrench	Handle
Tighten bolt	Transform
Put down wrench	Handle

The idea of subdividing jobs to eliminate non-value-adding labor did not, however, originate with Ford but often came from the workers themselves. Kipling (1897, website, Chapter 2) describes a fish-cleaning process on a trawler of the late nineteenth century:

> "Hi!" shouted Manuel, stooping to the fish, and bringing one up with a finger under its gill and a finger in its eyes. He laid it on the edge of the pen; the knife-blade glimmered with a sound of tearing, and the fish, slit from throat to vent, with a nick on either side of the neck, dropped at Long Jack's feet.
>
> "Hi!" said Long Jack, with a scoop of his mittened hand. The cod's liver dropped in the basket. Another wrench and scoop sent the head and offal flying, and the empty fish slid across to Uncle Salters, who snorted fiercely. There was another sound of tearing, the backbone flew over the bulwarks, and the fish, headless, gutted, and open, splashed in the tub, sending the salt water into Harvey's astonished mouth. After the first yell, the men were silent. The cod moved along as though they were alive, and long ere Harvey had ceased wondering at the miraculous dexterity of it all, his tub was full.

Each man doubtlessly knew how to do the entire job of cleaning a fish, but the complete task would have required him to pick up and put down a knife twice for each one. The job design was not perfect, as Manuel had to bend over for each fish while Long Jack apparently had to do the same, but the fact that the fish then "slid" to Uncle Salters suggests an efficient means of transportation similar to a rollway or work slide.

"Waste of the time of things" carries over into service activities such as airline travel. Boarding a large airplane through a single door translates into waste of the time of customers, which constitutes "waste of the time of things." The classification of customers as "things" seems unusual until we acknowledge them as sentient work in process to which cycle time is a CTQ characteristic. Inanimate widgets, unlike passengers, don't get angry if they have to wait longer than they should. People pay hundreds of dollars for airline tickets in exchange for shorter travel times, i.e., shorter cycle times, and anything that forces them to spend more time than necessary on air travel is an incentive to drive or take a bus instead. Long security lines, delays in delivery of baggage at the baggage claim area, and so on add about a mile per minute of delay to the distance customers are willing to drive instead of buying an airline ticket. "Waste of the time of people" applies, on the other hand, to inefficient activities by the airline's employees.

Waste of the Time of People

Physics defines work as the product of force and distance, so labor and exertion do not always translate into value-adding work. Atlas, the mythical Titan who carried the weight of the world on his shoulders, would have been surprised and outraged to discover that he had performed no work—at least not as defined by physics—whatsoever during the thousands of years he had been at this task. Frederick Winslow Taylor observed that many if not most steel mill workers used the same shovel to move materials ranging from iron ore to ash. A full load of ore was too heavy to move easily, but the worker put far more effort into moving the shovel than the load when he used the same shovel for ash. He went home tired and quite possibly exhausted from exertion even though he had done very little value-adding work. Taylor therefore recommended the obvious solution of using a narrow shovel for ore and a wide one for ash. This delivered far more work (in terms of value) for less physical effort. The concept carries over to the winter task of shoveling snow. A wide shovel is most labor-efficient for powdery snow, whereas a narrower shovel is best for wet snow or ice.

Waste of the time of people often but not always overlaps with waste of the time of things. Inventory stored on a shelf does not waste the time of people; but waste motion, a waste of the time of people, also wastes the time of things. This is the reason for SMED, which seeks to minimize setup times. Shigeo Shingo (Robinson, 1990, 333–334) pointed out that only the last turn on a bolt, the one that actually fastens two items together, adds value. If 15 turns are necessary to tighten the bolt, 14 of them, or 93 percent of the worker's effort, is waste motion that hides in plain view as part of the job. Shingo adds that the problem becomes even worse when the bolt's length is greater than the thickness of the part to be attached. Pear-shaped holes and U-shaped washers eliminate the problem so a bolt or nut can be tightened or released with a single turn.

Waiting for parts or tools at stockrooms is a waste of the time of people. If a $20-an-hour worker must spend half an hour to get a $5 part from a stockroom, the part costs the employer $15. This waste can be reduced enormously with, for example, tool vending machines that make tools along with consumable inventory available on the factory floor and at the point of use.

Being shot at is a powerful incentive to think innovatively, so military establishments were more than a century ahead of civilian enterprises in this kind of thinking. Practical breech-loading required the ability to close or open a cannon's breech with a single turn, or even a fractional turn, on the breech block. This was achieved with the interrupted thread breech plug, which could be pushed all the way into the breech after which a quarter turn engaged all the threads to seal the breech. The first such breech plugs had threading on only half their circumferences, which required a relatively long plug to contain the chamber pressures. The Welin breech plug of the late nineteenth century achieved threading on three-quarters of its circumference by placing the threads on segmented steps of varying radii (Brassey, 1899, 389). This increased its strength and made it even easier to handle. The key lesson here is that the ideal functional clamp (or equivalent) will require one or fewer turns with a wrench, handle, or other tool to do its job, and workers should recognize any task in which a bolt or nut seems to turn forever as wasteful of the time of people and possibly things.

This KPI does not apply to people who are idle for lack of work if the plant is running below capacity. Taylor pointed out the problem of soldiering, in which workers marked time (like soldiers marching in place) to deliberately limit productivity in admittedly justifiable retaliation for piece rate cuts by management. "Reverse soldiering," for want of a better term, takes place when dedicated and committed workers feel that they are not earning their pay unless they are always producing something. The something is often unnecessary inventory that ties up capital and also increases cycle time, i.e., adds to waste of the time of product.

"Waste of the time of people" also includes all non-value-adding activities as might be identified from a value stream analysis. Ford and Crowther (1922, 174) provide an example:

> We cut our office forces in halves and offered the office workers better jobs in the shops. We abolished every order blank and every form of statistics that did not directly aid in the production of a car.

This KPI, like the others (except possibly for waste of materials), applies to service as well as manufacturing industries.

Waste of the Time of People in Service Activities

It is generally useful, as shown for the airline example, to classify waste of the time of customers as "waste of the time of things"—in this case, sentient work in process to which cycle time is a CTQ characteristic. The difference between human work in process and inanimate work in process is that, in the latter case, the customer will notice only if the work is collectively late and does not meet a JIT delivery schedule. In the former case, every individual "work unit" notices delays and inefficiencies, which translate into external failure that might encourage the "work units" to look for ways to dispense with the inefficient service in question. Time wasted by employees is meanwhile "waste of the time of people" for which the organization must pay but that produces no value.

Consider the time a doctor's office spends in filing insurance claims, which adds no value for the patient. Time that nurses must spend walking in a hospital is yet another example of waste of the time of people, and the Henry and Clara Ford Hospital was designed explicitly to minimize the walking that nurses had to do during their work day. These wastes of the time of people contribute to excessive national medical costs.

Time spent in commuting to and from school, business meetings, and professional conferences also is waste of the time of people unless the travel time can be put to constructive use. Internet technology has recently offered ways to eliminate a considerable amount of this waste. Internet schools already exist, and they eliminate the need for school buses (waste of the time of the "product," in this case, students) and commuting by teachers (waste of the time of people). They also eliminate the waste of energy (fuel) associated with the latter activities, as well as the capital investment in the school along with its maintenance and upkeep cost.

The same concept applies to business meetings and professional conferences, some of which are already held on the Internet. If the interaction can take place online, it eliminates the need for travel (waste of the time of people and waste of energy) and overnight lodging (more waste of the time of people). This may not be good news for the airline and hospitality industries, but the poor service offered by the former simply accelerates the change by encouraging customers to look for ways to dispense with delays, added baggage fees, and poor quality service. Harrington (2005) elaborates considerably on the latter.

The hospitality industry must meanwhile realize that, no matter how good a hotel's service might be, technology can make anything obsolete.

The coaching or staging inn, which offered changes of horse along with overnight accommodations, was once an important part of the hospitality industry in an era in which tired horses had to be rested or exchanged relatively frequently, and 20 or 30 miles' progress was considered good for a single day. Truck stops now serve a similar purpose for commercial vehicles, as drivers can obtain lodging, food, and vehicle service in a single location, but driving intervals between stops are now measured in hundreds of miles. People will, of course, continue to need lodging for recreational travel, but the hospitality industry needs to realize that there is likely to be a declining demand for its business lodging and conferencing facilities, through no fault of its own.

Waste of Materials

Any purchased material that does not become part of the product is waste, unless it can be reused or sold as a by-product. Many consumable materials such as cutting fluids and solvents are necessary to the process but it is nonetheless necessary to recognize them as waste if they are discarded. Even if there are not environmental-specific disposal costs, their purchase cost might be avoidable or at least offset by salvage or recycling.

Ford, for example, converted slag from his blast furnaces into cement (*Ford News,* 1922, Vol. 3, 1). He also recovered paint solvents through adsorption so one gallon of purchased solvent did the work of 10 before it ceased to be serviceable. The ultimate, and apparently achievable by some factories, goal is a zero emission operation in which all purchased materials are reused or recycled into useful by-products.

Recycling and resale of waste products was not, however, enough for Ford, who described the ideal principle as follows (Ford and Crowther, 1926, 113). "Picking up and reclaiming the scrap left over after production is a public service, but planning so that there will be no scrap is a higher public service." The section on material and energy balances illustrates processes that generate huge piles of recyclable metal scrap, as did some of Ford's operations before he and his workers changed the job and/or product design to eliminate this waste. Arnold and Faurote (1915, 211–212) recognized this issue explicitly:

> The machine shop produces about 14,000 rings per day, 1 5/12 ounces each, say 1,240 pounds of finished rings from 13,000 pounds

> of ring stock, 11,760 pounds of stock, worth $294 wasted for the pleasure of cutting it into chips and using snap-ring piston packing.
>
> That is to say, $325 worth of ring-stock is supplied to the machine shop, $294 of this value is wasted, and $31 of stock value utilized in the finished work.
>
> These figures are not favorable to low-cost piston-ring production.

This excerpt illustrates the material and energy balance technique by accounting fully for the 13,000 pounds of ring stock (input) with outputs of 1,240 pounds of the desired product along with 11,760 pounds of chips. This gives the latter material waste absolutely no place to hide, which underscores the power of material and energy balances to meet any conceivable analytical needs of ISO 14001 and ISO 50001. The section on this technique shows, however, that it goes even further by accounting for cutting fluids and similar consumables as well. Redesign of the job or the part to reduce the necessary machining therefore reduces not only the waste of the metal but also the consumption of cutting fluid, which often has environmental disposal costs.

Waste of Energy

Any purchased energy (and this includes fuel for transportation) that does not transform the product is waste. This waste cannot be reduced to zero because lighting, heating, ventilation, and air conditioning are required to make the plant habitable, although Leadership in Energy and Environmental Design (LEED) can reduce these costs substantially.

It is possible to calculate for many processes, especially chemical processes and ore reduction operations, the amount of energy that is non-negotiable with the laws of nature to create the product. It is often possible to calculate this for machining operations as well; it takes a certain amount of energy to cut, pierce, weld, or drill the workpiece. This non-negotiable or theoretical energy requirement then allows a gap analysis against actual energy consumption by the process. This kind of gap analysis was in fact performed more than 100 years ago, and it underscores this book's contention that the United States does not have to invent or discover new solutions to its current economic problems. It need only implement the proven solutions that Harrington Emerson (1909), Frederick Winslow Taylor, Henry Ford,

Frank Gilbreth, and their contemporaries put on the shelf long ago for us to pick up and use whenever we want them.

Energy Efficiency Gap Analysis: Thought Process

Emerson (1912, Chapter 1, "Typical Inefficiencies and Their Significance," 4–5) provides several highly unfavorable comparisons of industrial processes to natural and biological processes. There may have been no practical alternative to the latter in 1909, but the comparison shows what *can* be done, and some of it *has* been done during the last decade. The key principle is that no improvement process can begin until the organization recognizes the opportunity for improvement, and that is where the gap analysis comes into play.

> An oil engine may reach 30 per cent thermal efficiency, but the salmon, assuming his whole weight to be pure oil, without consuming it, uses up several times more energy than is yielded by an equal weight of oil in combustion.
>
> The salmon uses atomic, not thermal, energy.

"Atomic" in this context does not mean nuclear-powered salmon but rather energy from biochemical processes.* The excerpt recognizes (1) the limitations of any thermodynamic power cycle that transforms heat into mechanical energy and (2) the fact that it is *possible* to get more useful energy from the fuel in question. Today's counterpart of Emerson's salmon is of course the fuel cell, which converts chemical energy directly into electricity. The reference continues,

> A filament, enclosed in a glass bulb, is heated to incandescence by an electric current and we use the glow for illumination. It takes a definite voltage and a certain number of amperes to heat a given filament to and keep it at the required brilliancy. There is frictional loss between lamp and dynamo, loss in the dynamo, losses in the steam engine driving the dynamo, losses in the boiler, in the furnace, in the transportation and mining of the coal.

---

* This reference does, however, hint at the prospect of nuclear energy more than 30 years before the first sustained fission chain reaction; it mentions the enormous amount of energy that is stored in the recently discovered element radium.

> Man wastes three-quarters of the coal in the ground, brings the remaining quarter to the surface by inefficient labor and appliances, doubles, trebles, or quadruples its cost by transportation charges to furnace door. Rarely is as much as 10 per cent of the energy of the coal transformed into electrical energy, and of this only 5 per cent can appear as light. Ten to twenty times as much light is provided as necessary on a writing table, because of the distance of the bulbs from the place where the light is needed. The light itself glows continuously, not only during intermittent work but often several hours before and after it is needed. Out of ten thousand B. t. u. in the coal mine we use in necessary light the equivalent of about six.
>
> The fire-fly converts the hydrocarbons of its food into light with an efficiency of 40 per cent. It flashes its light at intervals, thus making it most effective by contrast with the surrounding darkness, and it emits no more light than is necessary for its purpose.
>
> In production the fire-fly is about seven hundred and fifty times as efficient, in volume use ten times as economical, in time use twice as economical. The fire-fly is fifteen thousand times as efficient as his human rival.

It was not possible at the time to improve on the situation in question, but the comparison of the light bulb's efficiency to that of the firefly's bioluminescence again shows that it is *possible* to improve, and that the light bulb should therefore not be taken for granted. Today there are motion sensors that turn lights on when people enter a room and off when they leave, which addresses the issue of lights being on when they are not needed. The compact fluorescent light bulb and the light-emitting diode have meanwhile gone a long way toward turning electricity directly into light without wasting a portion as heat.

Emerson (1912, 9–10) continues by showing very explicitly the need for a *supply chain perspective* in the assessment of any activity's efficiency:

> If any human activity is followed out from initial reservoirs to final attainments, a similar sequence of losses will be found—losses gauged not by any ideal or unattainable standard, but by what is being continuously accomplished all around us. Even if, as yet, some of the high efficiencies seen in Nature are beyond reach, it is a greater reason for eliminating those wastes which are avoidable

> and which are primarily responsible for the starvation of men, women and children.

This excerpt also reinforces the desirability of always comparing what is being done against what can be done, i.e., performing a gap analysis, to expose waste that would otherwise easily be taken for granted.

With regard to the process itself, a material and energy balance is very useful for identifying material and energy wastes alike. This is a very useful analytical tool that supports ISO 50001 considerations like Energy Profile and Energy Baseline. It also ties in very closely with the Supplier, Input, Process, Output, Customer (SIPOC) model.

Material and Energy Balance

Material and energy balance is a chemical engineering technique that accounts for every resource that enters and leaves a process. Himmelblau (1967 and subsequent editions) provides an excellent textbook description, and the following statement is but one example of how it can reveal waste that would otherwise hide in plain view.

> Black smoke is unconsumed carbon—nascent heat—lost energy—wasted coal. A smoking chimney registers money lost (The System Company, 1911b, 28).

In this case, attention to the waste stream reveals what would today be not only an environmental aspect but also a waste of fuel and therefore money. This waste is easily visible to anybody who looks at the smokestack, but a material and energy balance can in fact identify all forms of material and energy waste. The standard approach begins with the steady state assumption.

Steady-State Assumption and Control Surface

Himmelblau (1967, 60) defines as a *system* "any arbitrary portion or whole of a process as set out specifically by the engineer for analysis. …note particularly the system boundary is formally circumscribed about the process itself to call attention to the importance of carefully delineating the system in each problem you work." This analytical system boundary is the control surface or control volume through which all inputs and outputs flow.

Most chemical engineering applications assume a steady state in which no material or energy accumulates or depletes inside the control volume. This means inputs must equal outputs just as debits must equal credits in accounting. Himmelblau (1967, 61) even presents a material and energy balance for a phenol manufacturing plant in ledger form. It is therefore technically impossible for any form of material or energy waste, and therefore for any ISO 14001 environmental aspect or ISO 50001-related energy loss, to hide from this analytical technique.

The analysis is generally far easier for most manufacturing operations than for chemical transformation processes in which stoichiometry (ratios of elements in the chemical compounds) is an issue. Carbon that enters a combustion process can leave as carbon dioxide, carbon monoxide, and/or even unburned carbon. In contrast, metal that enters as metal stock leaves either as product or machining chips.

Even a resource like paint does not undergo any chemical transformation that requires analytical consideration. If, for example, 200 pounds of pigment and 800 pounds of solvent go into a process, 200 pounds of pigment and 800 pounds of solvent must come out. If the 200 pounds of pigment are not on the product, they are probably part of the overspray and therefore waste. If the solvent evaporates to atmosphere, it is waste and possibly an environmental aspect. Shingo (Robinson, 1990, 101–102) and Ford and Crowther (1926, 67) addressed both of these wastes.

Application to Painting and Coating Operations

The Ford Motor Company had a coating process for cloth that used a mixture of solvents including ethyl acetate, alcohol, and benzene. An adsorption process captured the fumes in charcoal and, once the charcoal became saturated, hot steam drove the now-concentrated solvent vapors out to be condensed and collected for reuse. Ford and Crowther (1926, 67) reported that 90 percent of the solvent was recycled in this manner, so one gallon did the work of ten. It is of course necessary to weigh the capital and energy cost of the absorption-desorption cycle against the value of the solvent, but the process was clearly economical or Ford would not have used it. The key lesson is that most people would have focused only on the product (the coated cloth) and ignored the solvent. The material and energy balance makes the solvent clearly visible and forces people to think about it.

A client of Shigeo Shingo (Robinson, 1990, 101–102) painted fountain pen caps in a spray booth, and heavy rains sometimes caused the company's waste treatment plant to overflow and contaminate nearby rice paddies with paint solvents. This was a glaringly obvious "environmental aspect" and quite probably a violation of environmental regulations. Shingo looked at the paint booths and asked whether the operation's purpose was to paint the pen caps or the air. The plant manager understood immediately that much if not most of the paint missed the caps, and this was the source of his environmental problems. Replacement of the spray booth with nozzles that directed the paint directly at the pen caps eliminated the environmental problem along with the need to buy paint for the purpose of effectively throwing it away.

This problem, like many forms of waste, was obvious on sight *to somebody who knew how to recognize it*, but the fact that the pen company never noticed it reinforces the desirability of a formal material and energy balance on every operation.

Ledwith (2004) points out an identical or even worse problem in semiconductor manufacturing. The traditional spin-coating process, which deposits a photoresist or other coating on a silicon wafer and then spins the wafer to deliver a thin and uniform coating, *wastes 95 to 99 percent of the costly photoresist* and creates waste for whose disposal the company must pay. The waste stands in plain view to anybody who has ever seen a spin coater in operation, but only now are viable alternatives being developed to eliminate it.

Application to Steel and Aluminum Manufacture

Bennett (1951, 32–33) describes how Henry Ford recognized on sight waste that everybody else at his steel mill had apparently overlooked.

> One day when Mr. Ford and I were together he spotted some rust in the slag that ballasted the right of way of the D. T. & I [railroad]. This slag had been dumped there from our own furnaces.
>
> "You know," Mr. Ford said to me, "there's iron in that slag. You make the crane crews who put it out there sort it over, and take it back to the plant."

The permanent corrective action was to install powerful electromagnets to prevent iron particles from getting out of the blast furnaces in the first place. In this case the equivalent of "dumpster diving," or examination of what the process threw away as well as what it produced, revealed the waste to somebody who knew how to recognize it. A material balance would have worked equally well, as less iron came out of the steel mill than went in as iron ore. The discrepancy would have led to examination of the slag and discovery of the problem.

Energy balances also come into play in all processes that involve chemical reactions, and this includes the reduction of ores. Shreve and Brink (1977, 227) calculate that it requires 2.56 kilowatt-hours (kWh) to produce one pound of aluminum from alumina (Al_2O_3) at 1,000°C.

> In actual practice, however, some energy is used to bring reactants up to temperature and is lost in the sensible heat of the products. Some carbon monoxide is formed in the reaction, thus increasing the positive ΔH [enthalpy change], which amounts *practically* to between 6 and 9 kWh/lb. Consequently, this metal cannot be made economically unless low-priced electric energy is available.

The latter statement underscores the fundamental flaw in the argument that carbon taxes or cap and trade will encourage energy users to "be more efficient" because it is physically impossible to reduce aluminum from its ore for less than 2.56 kWh/lb. This number also, however, provides a theoretical best case against which the analyst can perform a gap analysis and then ask why the actual requirement is double or even triple the theoretical requirement. A Google search on "waste heat recovery" is highly instructive with regard to recovery and reuse of a portion of the sensible heat (energy changes caused by temperature differences in contrast to latent heat of phase changes, such as boiling or condensation). Carbon monoxide is, of course, an environmental aspect (pollutant), but it can also be sold as a chemical commodity.

Application to Machining Operations

Everything that goes into the operation, including not only stock but also cutting fluids and electricity, must balance everything that comes out, including not only the machined product but also the metal turnings or chips along with the spent cutting fluid (unless the latter can be reused). This is known as keeping one's eye on the hole (everything that is thrown away) as well as the doughnut (product).

As for chemical transformation processes, it is often possible to calculate the minimum energy that is necessary for a machining task. Kalpakjian

(1985, 477) defines the specific energy of a cutting process as "the total energy per volume of material removed." The power delivered to the work-piece is then F_cV, where F_c is the cutting force and V is the volume of material removed. Any disparity between this and the power actually drawn by the tool is therefore waste. Waste can occur, for example, if the tool must brake frequently, and regenerative braking can recover the wasted mechanical energy instead of dissipating it as heat.

The fact that cutting and grinding generate chips in the first place means, however, that the analyst must challenge the process even if the tool's power requirements approach the theoretical minimum. The chips are waste, which means the energy necessary to create them also might be waste. To put this another way, the entire cutting or grinding operation is waste if there is a more efficient way to do the job. Ford and Crowther (1930, 192–193) elaborate:

> A casting must be machined—sometimes by taking away thirty per cent of the metal; that is a waste.
>
> ...we wanted to forge, spin, or stamp parts and then build them into complete units by welding.

The reference adds that the forged or stamped parts were stronger and often lighter than a corresponding cast part, and it notes further that it was not possible at the time to detect casting defects. Nondestructive testing makes this possible today, but testing is at best a value-enabling or value-assisting operation; it does not actually add value to the product.

It is important to note here that material and energy wastes can sometimes serve as checks on one another. The analyst could conceivably overlook the machining process' power requirements if all or almost all of the power went into transformation of the product. The generation of the material waste itself is what provides a clue that the machining operation, which is value-adding by definition because it transforms the product, might not be the best way to do the job. This leads in turn to consideration of Design for Manufacture (DFM), Design for Assembly (DFA), and alternative processing methods. As but one example, if a large hole must be made in a piece of metal, less energy is required to cut through its circumference than to drill out the entire cross section even though the value added is the same in both cases.

Meanwhile, Mege (2000) shows just how much waste can be built into machining operations:

> The aerospace industry produces parts in quantities from a few units to thousands of pieces. They cut a huge amount of swarf from these parts—often, the weight of the finished part is only 15–20% of the original rough billet, so, on average, 80–85% of the aluminum is reduced to a heap of chips.
>
> The first U.S. manufacturer to buy one of our HAMs [High Agility Machines] four years ago posted a sign on his new "monster" reading: 'I eat 400 tons of aluminum per month!' The customer was forced to install a second swarf compactor, and now sells bricks of aluminum chips to recyclers.

The process in question therefore takes five or six aluminum billets and grinds all but one into scrap for recycling. When the workforce thinks in terms of "waste of the time of things, time of people, material, and energy," it will immediately recognize the aluminum chips as waste not only of material but also of the energy the tools required to grind most of the billets into chips. The mere presence of even a small pile of metal chips at the Ford Motor Company attracted immediate attention followed by efforts to redesign the part or the process to eliminate the need for machining. The material and energy balance, of course, forces this material and energy waste to become highly visible even if nobody notices the elephant in the living room or, in this case, the piles of aluminum in the shop.

Application to Power Generation

Consider a power plant whose input is coal and whose product is energy. The bottom line of the material and energy balance compares the energy contained in a pound of coal with the electricity that comes from the coal's combustion, and the difference is waste.

Some of the waste is unavoidable as stated by the Carnot cycle, which defines efficiency as the temperature difference between the hot and cold reservoirs (such as the furnace and the cooling water) divided by the absolute temperature of the hot reservoir. This is the bad news, and it gets even worse because the Carnot cycle is a theoretical best case that no real power generation cycle can achieve. This limitation is an incentive to (for example) react the coal with steam to produce hydrogen, which can then be used in a fuel cell. Fuel cells convert chemical energy directly into electricity and are therefore not subject to the limitations of the Carnot cycle, although their efficiency still falls short of 100 percent. The economic practicality of this change is, like most other capital investment decisions, subject to a net present value (NPV) analysis, but the visibility of the waste at least allows the power company to ask the right questions.

The material and energy balance also forces *all* other wastes, including the heat that the combustion gases carry up the smokestack, to become highly visible. Coal is the obvious input whereas carbon dioxide is the obvious output, but a rigorous material and energy balance requires consideration of the air as well. The combustion of a molecule of carbon (C) requires a molecule of oxygen (O_2), which, however, brings with it almost four molecules of nitrogen. Some of the heat of combustion that the power plant would like to go into the boiler must go into the carbon dioxide and nitrogen instead, which means it goes up the smokestack as waste. Matters become even worse in practice because excess air must be used to ensure complete combustion.

This waste, which the material and energy balance has made highly visible, suggests the obvious alternative of burning the coal in pure oxygen. There is then no nitrogen to carry sensible heat up the stack. No nitrogen oxides (pollutants) are generated, which means no pollution-control equipment must be installed to get rid of them. The combustion temperature is also higher, which yields a better Carnot cycle efficiency that translates into better efficiencies in real-world power-generation cycles.

The cost of pure oxygen (as supplied traditionally from cryogenic liquefaction processes) is an obvious obstacle to this agenda, but chemical looping combustion may overcome it. Chemical looping combustion involves the oxidation of a metal in air, with the oxidized metal then being reduced in the combustion process itself.

Material and Energy Balance, Summary

The effectiveness of the material and energy balance cannot be overemphasized because it is similar to cost accounting, in which debits must balance credits exactly. Cost-accounting techniques make it impossible for money to appear or vanish without identifying its source or destination. The material and energy balance makes it similarly impossible to purchase energy or materials without accounting for their ultimate disposition, whether as saleable product or waste. When this exercise is performed correctly, it is impossible for any material or energy waste to go undiscovered.

This book does *not* recommend the use of carbon emissions or a carbon footprint as a KPI. This metric's foundation is the controversial proposition that costly and extraordinary measures must be taken to reduce global carbon emissions. "Extraordinary" means measures beyond those justified by engineering economics to reduce any waste of energy, whether from carbon or noncarbon sources. It means specifically carbon sequestration, selection of noncarbon energy sources, or anything else that cannot pass a standalone discounted cash flow or NPV analysis.

This next section addresses this controversial issue in detail because some supply chain partners are now requesting carbon footprint information from their suppliers. This, in contrast to good engineering and manufacturing practices whose purpose is to reduce energy waste in general, is not an effective use of supply chain resources. The next section will equip this book's users to educate management, supply chain partners, and the public as to why expenditures on the climate change agenda are simply another form of waste.

Do Not Use Carbon Emission Metrics

An organization or supply chain that measures its carbon dioxide emissions in response to political rhetoric about global warming is doing the wrong thing for the wrong reason. If it measures these emissions to identify and quantify energy usage and therefore possibly waste, it is doing the wrong thing for the right reason because energy wastes from noncarbon sources are invisible to this metric. No energy waste whatsoever can, on the other hand, hide from the recommended KPI, which is simply "waste of energy." Any action to reduce it will have the incidental effect of reducing carbon emissions if the energy in question happens to be from fossil fuel sources.

Meanwhile, there are far too many special interests behind initiatives like carbon taxes and cap and trade to make the climate change agenda trustworthy. Suppliers therefore need to remind customers who ask for carbon emission metrics that social responsibility—the argument often used by cap-and-trade advocates—consists very simply of the obligations in Table 2.3.

Ford proved unequivocally that all these objectives are synergistic and mutually supporting instead of antagonistic. A business can indeed charge low prices, pay high wages, and make money hand over fist by removing waste, including environmental waste, from the supply chain that delivers the product.

In the absence of clear and convincing proof that carbon dioxide endangers public health and safety, extraordinary efforts to suppress its emission

Table 2.3 Social Responsibility Obligations

1. Pay the highest possible wages to employees. Ford and Crowther (1922, 117) elaborate, "It ought to be the employer's ambition, as leader, to pay better wages than any similar line of business, and it ought to be the workman's ambition to make this possible," and this single sentence also applies to LMS:2012's provision for management and workforce commitment.
2. Charge the lowest possible prices to customers, as opposed to what the market will bear.
3. Pay the highest possible compensation to suppliers while keeping in mind their reciprocal obligation (item #2) to deliver the lowest possible costs. This does not mean giving the store away, and zero base pricing (ZBP) means zero tolerance for supplier wastes and inefficiencies. A socially responsible business does not, however, try to squeeze unrealistic price concessions from suppliers that compel the suppliers to cut wages, take shortcuts on quality, or shortchange their own investors.
4. Pay the highest possible returns to investors who have entrusted the organization with their capital.
5. Protect public and employee health and safety from pollutants and workplace hazards, respectively. No supply chain has the right to discharge toxic pollutants into the air or water where they will endanger people and wildlife downstream, but the classification of a gas that all animals exhale and all green plants consume as a threat to public safety requires considerable scrutiny.

are socially *irresponsible* because one or more stakeholders in the supply chain—customers, suppliers, employees, and/or investors—have to pay for them. No agenda that raises the cost of a basic necessity like food through carbon taxes, cap and trade, or corn ethanol subsidies can claim to be socially responsible. The same goes for an agenda that raises the cost of energy to heat, cool, and illuminate homes or to transport people to and from their jobs. The destruction of working people's jobs through higher energy costs that encourage manufacturers to move offshore is anything but socially responsible.

Meanwhile, this chapter has defined "extraordinary" as any expenditure that cannot meet the criteria of discounted cash flow analysis, and Equation (2.3) shows the general procedure. "Discounted cash flow" means that future cash flows, whether positive or negative, are discounted for the time value of money. If, for example, the required rate of return is 10 percent, $100 a year from now is worth only $90.91 today. A benefit such as revenue or cost avoidance is positive whereas an expenditure or cost is negative, and this applies to all terms in the equation. An investment is acceptable if the NPV is positive and unacceptable if it is negative.

Equation (2.3): Net Present Value

$$NPV = P + \sum_{k=1}^{N} \frac{A_k}{(1+i)^k} + \frac{S}{(1+n)^N} \tag{2.3}$$

where

- N = life of the investment or project (generally years).
- i = required rate of return expressed as a fraction, e.g., 8% = 0.08.
- P = present outlay or investment. This is generally negative because the money must be paid up front.
- A_k = income at the end of period k. It is positive for revenues and savings and negative for costs. It is often convenient to deduct the operating or maintenance cost from the income to get a single number. This term can also include the tax benefit of depreciation.
- S = salvage from sale of asset at the end of the project or investment life. S is positive if the item can be sold and negative if there is a disposal cost. This term is generally insignificant for investments or projects with long time frames.

The salvage term, which is simply the present value of a single future income or payment, also puts the cap-and-trade agenda's mantra of "pay now or pay later" in an entirely new and highly unfavorable light. This is obvious if S is the cost of mitigating climate change effects if and when they occur 50 or 100 years in the future, whereas A is the annual cost of quite potentially futile attempts to prevent those changes. It should be noted similarly that the recycling value (or disposal cost) of a capital asset is a relatively minor consideration because the salvage term is usually the least important one in the equation.

The purchase of a hybrid electric vehicle might therefore easily be socially responsible for delivery operations that require frequent braking, which the driver should recognize as "waste of energy." The discounted value of the fuel thus saved (A) is quite likely to exceed the additional present cost of the hybrid option (P). Recovery of the braking energy should therefore translate into lower costs for the customers, higher profits for the company, and higher wages for the driver.

The limited ability of alternative energy sources like solar cells and wind turbines to sell themselves is, on the other hand, prima facie evidence that they often cannot pass an NPV analysis. Note, however, that the engineering or managerial economics depends on the location of the proposed investment; the NPV for solar power will be far better in the Southwest than in New England. Meanwhile, the relative strength of available wind will affect the NPV for a wind turbine.

It is the job of those who wish to sell alternative energy to get its price below that of traditional energy from sources like coal. Meanwhile, the fact that many renewable energy suppliers are doing their best to achieve this is an incentive for power companies that rely on coal to burn it more efficiently or, noting again the limits of thermodynamic power cycles, look more deeply into fuel cell technologies. Free market competition of this nature is how technological progress takes place.

The next section will examine closely the proposition that (1) climate change is a problem and (2) whether extraordinary measures to prevent it will even be effective.

Is Climate Change a Problem?

"Climate change denial" is a disingenuous phrase that dismisses opponents of carbon emission taxes, cap-and-trade mandates, and so on as ignorant, unscientific, and/or dishonest. No educated person denies that climate change is part of Earth's geologic history. Fossils of fish and marine animals can be found in what are now the Great Plains, which were the bed of the Western Interior Seaway during the Cretaceous Period. There were no polar ice caps at the time. Had humans existed, they could have sailed from the Hudson Bay to the Gulf of Mexico through what is now the Midwest. New York's Finger Lakes and Pennsylvania's Pocono Mountains are, on the other hand, legacies of a time when glaciers extended south of Canada and New England, and there is no guarantee that the last Ice Age was in fact the last. The advance and retreat of the glaciers took place without anthropogenic greenhouse gas emissions, unless one counts the campfires that prehistoric humans lit to stay warm and keep various animals away.

Proponents of carbon taxes and cap-and-trade mandates often argue that we must either pay to prevent climate change now or pay for its consequences later. The concept of discounted cash flow has already shown that "pay later" is almost universally superior to "pay now." Second, the anticarbon camp fails to acknowledge that their approach could easily turn out as "pay now *and* pay later" when climate change happens anyway. What opponents of carbon taxes and cap-and-trade mandates deny is the proposition that displacement of trillions of dollars in economic resources will deliver commensurate returns to the human condition anywhere in the world.

Bryant (2004, webpage) reports, for example, that the Maldives could become uninhabitable in 100 years. "No wonder it was the first country to sign up to the Kyoto Protocol, which sets targets for cuts in industrialized countries' greenhouse gas emissions." This simplistic approach will, however, place the inhabitants of the Maldives at far greater risk if the industrialized world has the poor judgment to act on it. Compliance with Kyoto or its equivalent will waste resources that might later be needed to evacuate the Maldives' people and supply them with new homes when, as is quite likely, *greenhouse gas mitigation doesn't prevent the climate change in question.* Sea levels have been rising since the end of the Ice Age, and any homes that prehistoric humans built in what is now the English Channel have been under water for thousands of years. If New York (or New Amsterdam) had been founded a few thousand years earlier, the scene in the Steven Spielberg movie *AI* in which some of the city's skyscrapers are under water might be reality.

The equation for NPV, and specifically the salvage term at the end of the project's or asset's life, also is very useful for assessing the situation of the Maldives. Suppose that the required rate of return on investments is only 5 percent, noting the negligible interest rates that currently prevail. (Twenty percent was common in textbook examples from the 1970s and 1980s.) The net present cost of the contingency of evacuating the Maldives 100 years from now and buying new homes for its people is then 0.76 percent of whatever that future outlay may be.

Adaptation to climate change is therefore far cheaper than multitrillion dollar efforts to emulate King Canute's futile command that the tide not come in. Canute's honesty is legendary because he, unlike many backers of the cap-and-trade agenda, not only admitted but proved deliberately that human governments cannot command nature. The effectiveness of adaptation, unlike that of Kyoto or similar legislation, can be guaranteed. It is highly unlikely that the cap-and-trade agenda will save the Maldives, or for that matter coastal properties, from whatever the world's climate decides to do. It is a certainty, however, that migration to greener pastures, as humans and animals have done throughout history, will protect the lives and well-being of all the people involved. The decision to not squander limited financial and other resources on the climate change agenda will help ensure that national or international resources are available to provide the necessary migration if and when it becomes necessary.

Another inconvenient truth the advocates of cap and trade leave out is the undeniable fact that poverty kills people through inadequate nutrition, inadequate sanitation, and lack of access to first-class medical care. Cost-ineffective carbon emission mandates encourage employers to close energy-intensive businesses or move them, along with all their carbon emissions, offshore. In other words, they don't even get rid of the carbon dioxide; they just get rid of the jobs. Fossil fuel taxes or mandate costs increase the prices of necessities like heat, electricity, air conditioning in warm climates, and transportation including to and from a workplace. They also raise the cost of food because of its transportation cost, and the politically correct "alternative energy" agenda has already caused hunger among the world's poor. As reported by O'Grady (2007, A16),

> The sharp increase in Mexican corn prices, which fueled the tortilla price spike, followed big price increases for corn on international markets over the past year. The main cause, according to most commodity analysts, was the U.S. decision to subsidize ethanol made from corn. Growers who previously marketed their harvests to food and livestock companies suddenly have new demand from ethanol producers, who are also armed with a subsidy to make their bids more attractive. The increase in demand from government-subsidized ethanol producers pushed up prices.

Rattner (2011, webpage) adds of ethanol subsidies,

> But rather than ameliorate the problem [of rising corn prices], the government has exacerbated it, reducing food supply to a hungry world. Thanks to Washington, 4 of every 10 ears of corn grown in America—the source of 40 percent of the world's production—are shunted into ethanol, a gasoline substitute that imperceptibly nicks our energy problem. Larded onto that are $11 billion a year of government subsidies to the corn complex.

There is nothing socially responsible about "reducing food supply to a hungry world," a problem that carbon taxes and cap and trade can only exacerbate. For every person (if any) this agenda saves from nebulous threats like rising sea levels, weather changes, and so on, it is likely to kill dozens if not hundreds through a far lower standard of living if not outright poverty. Closer inspection of the agenda in question shows that social responsibility has very little to do with it.

Special Interests and the Climate Agenda

Businesses have both a legal and ethical obligation to provide safe work environments for employees and not release genuine pollutants that endanger human or by implication animal health. Confusion of public safety with individual or corporate financial interests is, on the other hand, commonly known as a conflict of interest. Suppose for example, that Henry Ford had, instead of working to make the Model T so inexpensive that almost everybody could afford it, proclaimed that equine emissions were a public health hazard. This claim would have had far more merit in terms of certainty than claims about carbon dioxide emissions because horse waste is obviously a sanitation problem.

Ford's next step might have been to join other automakers in calling for taxes on horse food, which would have made ownership of horses far more expensive and encouraged people to buy automobiles instead. It could have then been rightly argued that Ford and his fellow automakers had confused the public safety with their economic self-interests, especially since the exhaust from their products also was a genuine pollutant, even though the implications were not fully recognized at the time. This is not, of course, what Ford did, and he would have denounced in the harshest terms any business that tried to achieve through legislation what it could not or would not do through workmanship, engineering, and value for customers.

As with the first mass-produced automobiles, certain forms of greenhouse gas mitigation are not without environmental consequences either. Sequestered carbon dioxide can increase the acidity of groundwater, which can leach various undesirable chemicals into what will eventually be used as drinking water. Carbon sequestration also has some of the environmental drawbacks involved in the fracking (hydraulic fracturing) process for natural gas extraction. Many people in the Northeast, especially Pennsylvania and New York, object to the use of fracking on the Marcellus Shale formation despite the economic benefits because of what they are afraid it might do to the local aquifers. This reinforces the need to question whether the purpose of the cap-and-trade agenda is really to protect the public safety or to enrich special interests. Senator Kirsten Gillibrand (2009, webpage) identified some of those special interests inadvertently when she proclaimed how good cap and trade would be for New York's financial sector.

> According to financial experts, carbon permits could quickly become the world's largest commodities market, growing to as much as $3 trillion by 2020 from just over $100 billion today. With thousands of firms and energy producers buying and selling permits to emit carbon, transaction fees for exchanges and clearing alone could top nearly half a billion dollars.
>
> ...An infrastructure is already beginning to form, as entities like the New York Stock Exchange, J.P. Morgan Chase, Goldman Sachs, and the new Green Exchange are developing carbon trading platforms or expanding their environmental trading desks.

Strassel (2007, A10) adds,

> There was a time when the financial press understood that companies exist to make money. And it happens that the cap-and-trade climate program these 10 jolly green giants are now calling for is a regulatory device designed to financially reward companies that reduce CO2 emissions, and punish those that don't.
>
> Four of the affiliates [U.S. Climate Action Partnership]—Duke, PG&E, FPL and PNM Resources—are utilities that have made big bets on wind, hydroelectric and nuclear power. So a Kyoto program would reward them for simply enacting their business plan, and simultaneously sock it to their competitors.

Strassel also names the now-defunct Lehman Brothers as a beneficiary of cap and trade along with General Electric, which stood to profit from the sale of wind turbines and solar equipment. O'Keefe (2009, webpage) cites Enron, which was hardly a model for social responsibility, as another self-serving agitator for cap and trade:

> The future Enrons and Bernie Madoffs of the world would like nothing better than to see the U.S. impose a new market for carbon emission trading.

> …The cap-and-trade system being touted on Capitol Hill would create a multibillion-dollar playground that would, once again, create a group of wealthy traders benefiting at the expense of millions of average families—middle to low-income households that would end up paying more for food, energy, and almost everything else they buy.

"California's Biofuel Rules Rejected by Judge" (2011) adds that a federal judge believed that the purpose of California's mandate for low-carbon fuels had more to do with enriching in-state biofuel producers than genuine protection of the environment. The California law included a preference for in-state ethanol over ethanol from other states, ostensibly because of the greenhouse gases produced during transportation of the non-California biofuels. This justification did not fool the judge and it should not fool anybody else.

None of this discussion challenges the impartial findings of scientific research that concludes that carbon dioxide is (or is not) a contributor to global warming. It shows, however, that there is far more than impartial science behind the cap-and-trade agenda, which is why Lean implementers need to question it and encourage their stakeholders to do the same.

LMS:2012 includes provisions in which managers and executives must demonstrate commitment to the Lean program, and this includes matching actions to words or "walking the talk." This leadership behavior is conspicuously absent among proponents of greenhouse gas emission reduction.

Cap-and-Trade Community Doesn't Walk Its Talk

The title of an article by Gilligan (2009) speaks for itself: "Copenhagen Climate Summit: 1,200 Limos, 140 Private Planes and Caviar Wedges:— Copenhagen is preparing for the climate change summit that will produce as much carbon dioxide as a town the size of Middlesbrough." The article adds that "The top hotels—all fully booked at £650 a night—are readying their Climate Convention menus of (no doubt sustainable) scallops, foie gras and sculpted caviar wedges." £650 is roughly $1,000 depending on the current rate of exchange, and it is doubtful that most manufacturing professionals' employers would allow them to spend a fraction of this on lodging even in large metropolitan areas. The limousines and private planes, both of which emit far more than each passenger's fair share of carbon dioxide, meanwhile speak for themselves.

China, which is now the world's largest manufacturer—and therefore soon to become the largest military power—has meanwhile committed explicitly to the reduction of the carbon intensity of its operations (BBC News, 2009). This is emphatically not a commitment to reduce its overall greenhouse gas emissions but only to reduce them per unit of gross domestic product.

> Carbon intensity, China's preferred measurement, is the amount of carbon dioxide emitted for each unit of GDP.
>
> ...our Beijing correspondent says this is a commitment to make Chinese factories and power plants use fuel more efficiently and get better results.

Carbon intensity is indeed China's preferred measurement, and rightly so. It's good public relations for those who think a 40 percent reduction in carbon intensity translates into a 40 percent reduction in greenhouse gas emissions, but this "commitment" is nothing of the sort. If China cuts its carbon intensity 40 percent and doubles its GDP by 2020, its carbon dioxide emissions will increase 20 percent. The bottom line is that *China has committed only to the commonsense objective of eliminating waste of energy from its operations and therefore to the same KPI this book recommends.*

Table 2.4 summarizes the key talking points that organizations can share with supply chain partners and also the general public with regard to this issue and also to answer any ISO 50001-related questions about greenhouse gas emissions. The addition of the last item can turn the table's contents into a good corporate policy.

The next section will show that the four recommended KPIs should identify all forms of waste in any operating process. They also cover the recommended KPIs of Goldratt's Theory of Constraints, the Seven Wastes of the Toyota Production System, and the purpose of almost every Lean manufacturing technique.

Recommended KPIs Identify All Operating Wastes

The four recommended Lean KPIs—waste of the time of things, waste of the time of people, waste of materials, and waste of energy—are sufficient to identify comprehensively all wastes in an operating process. The word *operating* is important because idle or obsolete assets that tie up capital cannot

Table 2.4 Climate Change Talking Points and Sample Company Policy

XYZ Corporation defines social responsibility as a square deal for all supply chain stakeholders including customers, suppliers, employees, and investors along with workplace and environmental safety.

- It is the policy of XYZ Corporation to eliminate all forms of waste from its activities, including wastes of time, materials, and energy, to provide more value for our employees, investors, and supply chain partners.
- Any economically viable action that reduces the waste of energy will have the incidental effect of reducing carbon emissions if the energy is from fossil fuel sources. XYZ Corporation is fully committed to this course of action.
- XYZ Corporation does not, however, acknowledge as a workplace or environmental safety problem a gas that all animals exhale, and all green plants consume. The reasoning is covered below.

1. Climate change is a proven fact of Earth's geologic history. What remains to be proven is whether:
    - Costly efforts to reduce or sequester carbon dioxide emissions can prevent or even mitigate climate change, noting the extremes of which the Earth has proven itself capable in the complete absence of industrialized societies.
    - Any mitigation so achieved will be worth even a small fraction of the money necessary to achieve it, and that adaptation to the change (with national or international assistance as needed) will be far more cost effective.
2. When an entity confuses public health, safety, and welfare with its own economic welfare, it has a conflict of interest in the issue under discussion. Dozens of corporations and nonprofit organizations have multibillion dollar stakes in cap and trade, and this needs to be recognized.
3. The actions of the foremost advocates of controls on carbon emissions show that they do not take the matter seriously, at least when they think nobody is watching them. Those who are not willing to lead by example should not expect others to follow them.
    - Note also that China has committed only to reduction of the carbon intensity of its operations, which is simply good engineering and not a commitment to reduce overall greenhouse gas emissions.
4. The cost of any form of waste, including efforts to measure carbon emissions as opposed to energy costs in general, the purchase of carbon offset credits, or investment in costly noncarbon energy sources, must be passed on to one or more of these stakeholders. This company will therefore not charge higher prices, pay lower wages, or shortchange investors and suppliers to support this questionable agenda. We must accordingly insist that the costs of any such activities be borne entirely by those customers that require them.

generate any of the wastes in question, but LMS:2012 (like official standards for quality, environmental, and energy management) focuses on operating processes.

Lean KPIs and Goldratt's Theory of Constraints

In the Theory of Constraints (TOC), throughput (product ready to ship and with customers who want it), inventory, and operating expense provide everything a manager needs to know about the health of a production control system. Taiichi Ohno (1988, ix) adds lead time or cycle time as a KPI: "All we are doing is looking at the time line from the moment the customer gives us an order to the point when we collect the cash. And we are reducing that time line by removing the non-value-added wastes."

Lead and cycle time are similar in concept, but there is a technical difference: lead time is the time between placement and fulfillment of an order, whereas cycle time is the time between the actual start and completion of the work. They are identical in a perfect JIT system, i.e., the work begins the instant the order is placed. Lead time is otherwise always greater than cycle time unless the supplier carries an inventory of finished goods that can then be replenished at leisure. The latter approach is of course inconsistent with modern thinking.

Cycle time, throughput, and inventory are, however, redundant because any two define the third via Little's Law: cycle time = inventory divided by throughput. This returns us to three independent KPIs of the production system itself:

- Cycle time, inventory, and operating expense
- Throughput, inventory, and operating expense (Goldratt's TOC)
- Cycle time, throughput, and operating expense

All are, however, direct functions of the Lean KPIs in Table 1.1:

- Cycle time and inventory are both direct functions of waste of the time of products. Reliance on this single KPI leaves throughput as an unknown in Little's Law, but minimization of this KPI means the inventory will be the minimum possible for any given level of throughput. Meanwhile, Ford's statement "Not one hour of yesterday, nor one hour of today can be bought back" ties in directly with Goldratt's observation that time lost at the constraint is lost forever.

Lean KPIs and the Toyota Production System

The TPS defines Seven Wastes, but these are in fact measurable by the four recommended KPIs as shown by Table 2.5.

Table 2.5 TPS Seven Wastes as Functions of Four Lean KPIs

1. Overproduction and consequent inventory generation is reflected by the waste of the time of the product.
2. Waiting is encompassed by waste of time of products and people.
3. Transportation in terms of both time and cost is another TPS waste. The two factors are related but not totally dependent. A shipment of green wood may arrive as quickly as one of dry wood, but it is more expensive because the water in the wood has no value. Ford insisted on the shipment of dry wood, and preferably wooden parts, for this reason. Unnecessary or inefficient transportation is reflected by waste of energy (fuel) along with the waste of the time of product ("float" or products in transit).
4. Processing wastes consist of waste of the time of product (e.g., setup, which is reduced by single minute exchange of die), material, energy, and quite likely the time of people.
5. Inventory is, as shown above, a direct function of throughput and cycle time per Little's Law. As shown for TOC metrics, minimizing waste of the time of product automatically minimizes cycle time and inventory for any given level of throughput.
6. Waste motion is waste of the time of people.
7. The cost of poor quality will include all four wastes, including, for example, time, material, and energy necessary to replace scrap or rework nonconforming parts.
a. The four Lean KPIs therefore encompass the cost of poor quality, whereas it is primarily the job of the QMS to address it.
b. The four Lean KPIs also, however, identify wastes that are invisible to the QMS. The QMS is designed to focus on the traditional costs of poor quality: prevention, measurements and inspection to screen out nonconforming work (appraisal), rework and scrap (internal failure), and nonconforming work that reaches the customer (external failure). A process can, however, waste enormous quantities of time, material, and energy even though it delivers no nonconforming work whatsoever. Operating expense, or at least the part that can be avoided, is a function of the Toyota Production System's Seven Wastes (Ohno, 1988, 19–20). These also are direct functions of the four recommended Lean KPIs.

This chapter has shown so far that the four recommended Lean KPIs cover the principal metrics of both TOC and TPS. It will now show that they also cover the goals and objectives of most Lean manufacturing techniques.

Lean KPIs and Lean Manufacturing Techniques

Table 2.6 lists major Lean manufacturing and Lean enterprise techniques, along with their relationship to the four Lean KPIs.

Consider, for example, a plastic container of soybean or whey protein that is anywhere from 33 to 40 percent empty as delivered, even though the weight of the contents is as advertised. This is easily recognizable as waste of materials, specifically the extra plastic, and waste of energy. Reduction of the container's size allows shipment of 50 percent more containers in a truck, which reduces the per-unit shipping cost as reflected by energy (fuel) consumption. If the organization regards it as good public relations to talk about its carbon footprint, the carbon emissions per delivered container also are less. The shipment of air, whether in packing materials, partially full containers, or the product itself, should be avoided if possible.

The indicated four KPIs do not account explicitly for the fact that the oversized soy powder containers also occupy 50 percent or more of the retailer's shelf space than they need to, and efficient shelf space utilization is obviously a CTQ characteristic for retailers. This leads to consideration of another possible KPI: "waste of capital investment."

Waste of Capital Investment

The "waste of energy" KPI is sufficient to identify the problem with soy powder containers that contain up to 40 percent air when full. To this, "waste of the time of people" may be added if forklift and truck drivers must move air instead of product. As stated above, however, retail shelf space is valuable, and the retailer earns no more money from stocking air than a production worker earns by walking. The inefficient containers therefore waste part of the retailer's capital investment, and the waste can be detected from that perspective as well. The retailer is in fact more likely to notice waste of shelf space than unnecessarily high shipping costs due to waste of energy. It is first, however, necessary to break waste of capital investment down into the categories of non-operating and operating processes.

Table 2.6 Lean KPIs and Lean Enterprise Methods

• 5S-CANDO (Clearing up, Arranging, Neatness, Discipline, Ongoing improvement) reduces waste of the time of people (e.g., due to the need to search for tools instead of having them available in a specific place) and also helps expose problems that could result in equipment downtime (waste of the time of people and possibly things, if the product or service has to wait) or quality problems.
• Autonomation (jidoka), error proofing (poka-yoke), and source inspection all prevent internal failure (rework or scrap). The cost of rework or scrap includes the time of things, time of people, materials, and/or energy necessary to make good the losses.
• DFM, DFA, and robust design offer several benefits:
• Lower material and/or energy requirements, e.g., designs that require less machining
• Ease of assembly, and therefore less waste of the time of people
• Robustness to process variation, which results in less rework and scrap
• Higher process capabilities due to more forgiving designs, which also results in less rework and scrap
• JIT, Drum-Buffer-Rope, kanban, and other pull production systems reduce inventory and therefore waste of the time of things. The same is true of heijunka or production leveling.
• Preventive maintenance prevents equipment breakdowns and consequent waste of the time of people and possibly things.
• SMED reduces non-value-adding setup time and thus facilitates small-lot or single-unit processing. This reduces cycle time, and therefore waste of the time of things, and also waste of the time of people on labor-intensive setup activities.

Non-Operating Processes and White Elephants

Obsolete or unused equipment, buildings, and similar assets cannot waste the time of people or things, as they are not part of any process. Idle resources do not use materials or energy and can therefore waste neither. Ford and Crowther (1922, 174) stated explicitly, however, that "everything and everybody must produce or get out." Anything that is not part of a process and is unlikely to become part of a process represents idle capital that can be put to better uses. If the organization has no further use for a machine tool, it should sell it to somebody who can put it to work. "White elephant" industrial equipment is in fact often put up for sale on eBay. The owner of idle buildings and unused land must meanwhile pay to insure them (even if only for liability) and possibly maintain them. These also should be sold if there are no foreseeable uses for them.

LMS:2012 does not, however, recommend addition of this metric to the basic set of KPIs because it rarely applies to assessment of operating processes. Remember that "waste of energy" was adequate to detect the partially filled containers that wasted retail shelf space because the containers were part of a process. "Waste of the time of things" meanwhile exposes inventory and therefore capital tied up in inventory. "Waste of capital investment" therefore applies mostly to assets that are not part of any process. It is very useful to be aware of this issue and to sell for whatever one can get any asset that is no longer productive, but this is generally an offline consideration as opposed to a process consideration. The next section will, however, treat applications in which operating processes waste the time of capital assets.

Waste of Capital Assets in Operating Processes

The production of inventory to absorb overhead and/or capital depreciation costs is a dysfunctional value-destroying activity. These costs are sunk costs, which means generation of unnecessary inventory will not make them go away; it simply ties up even more cash in a form that cannot be readily used. This chapter has already categorized idle or underutilized capital assets *for which value-adding work is available* as "waste of the time of things." The concept applies directly to the constraint or capacity-constraining resource in TOC and to other assets as well.

Chapter 3 includes examples of tradeoffs between fuel efficiency and transportation speeds for trucks, ships, and similar assets. Operation at lower speed reduces the per-load fuel cost but also reduces the loads the asset can move per month or per year. Ford and Crowther (1926, 119–121) justified payment of high wages for Ford's Great Lakes ore carriers because "the important thing is to see that you get the full use out of the big investment, which is the ship." Ford's concern was that the ship be kept in motion instead of spending a long time in port, which was likely to happen if anyone but the best sailors were in charge of it.

Operation at lower speeds to reduce fuel costs means it is necessary to buy more ships to move the same overall load. It is necessary at this point to perform an economic assessment of the tradeoff between fuel savings and the higher capital investment *unless there is no paying work for the extra ships (or trucks).* If, on the other hand, there is not enough work to keep an existing fleet busy, it may make sense to operate at lower speeds. The trade-off is then entirely between lower fuel costs and longer delivery times plus higher per-mile labor costs.

Chapter 3 also includes an example of the application of NASCAR pit crew techniques to reduce turnaround times for commercial aircraft. The idea here is to get more use out of the airplane because the incremental profit from even one additional flight per day is pure profit. If paying work is available for the aircraft, any time it spends elsewhere but in the air constitutes "waste of the time of things."

Summary: Lean Key Performance Indicators

Three of the four recommended KPIs come with rigorous analytical techniques—specifically cycle time accounting for waste of the time of things, and the material and energy balance for wastes of material and energy—that give the indicated wastes *absolutely no place to hide.* Meanwhile, traditional industrial engineering and human factors techniques should identify waste of the time of people, although many job designs are still unfortunately very wasteful of human labor.

There are also cases in which the waste of the time of capital assets *for which paying work is available* should be taken into account. It is important to distinguish clearly between these situations and the dysfunctional objective of producing excess inventory to "absorb overhead" or "keep assets busy."

The next chapter will apply the four recommended KPIs into an integrated Lean assessment procedure and continuous improvement cycle.

Chapter 3

Integrated Lean Assessment

This chapter offers an integrated Lean assessment program that compels all forms of waste to become visible and obvious. This supports ISO 14001 and ISO 50001 directly and may, in fact, be the only approach that is necessary to meet or exceed the analytical requirements of these standards. The continuous improvement cycle of Isolate, Measure, Assess, Improve, and Standardize (IMAIS) uses the recommended Lean key performance indicators (KPIs) to identify waste proactively and then remove it.

All continuous improvement processes are variants of Plan, Do, Check, Act (PDCA) or Plan, Do, Study, Act (PDSA). These include not only the Ford Motor Company's Team-Oriented Problem Solving, Eight Disciplines (TOPS-8D) and the Automotive Industry Action Group's similar Continuous Quality Improvement (CQI-10) but also Six Sigma's Define, Measure, Analyze, Improve, and Control (DMAIC) and the Air Force's Observation, Orientation, Decision, Action (OODA) loop. The basic idea is to (1) identify or define a problem or opportunity, (2) conduct an assessment or analysis to obtain a potential solution, (3) test the solution and, if it works, (4) make it the new standard or "best known way" for the job in question.

Why IMAIS?

Why introduce yet another continuous improvement cycle when so many are already available? An 8D, CQI-10, or DMAIC project usually begins when the organization realizes there is a problem or, in proactive contexts, an opportunity. IMAIS sets out, on the other hand, to find the problems and

Table 3.1 IMAIS

1. Isolate a process (using the process perspective including Supplier, Input, Process, Output, Customer [SIPOC]) with a control surface or control envelope for analytical purposes.
2. Measure the process's efficiency in terms of the four recommended KPIs (or their equivalent): waste of the time of things, time of people, materials, and energy.
3. Assess the process for the root causes of the wastes in question, and determine the economic practicality of their removal.
4. Improve the process, and verify the effectiveness of the improvements.
5. Standardize the improvements to make them permanent, and deploy them to related activities through best practice deployment, AIAG's (2006, 137) Lessons Learned data base, and AIAG's (2006, 172–173) Read Across/ Replicate Process.

opportunities proactively, and it does this with quantitative methods that make it almost impossible for them to hide. It then continues with the standard improvement cycle that identifies a root cause, implements an improvement, verifies that the change works, and makes the improvement the new standard for the activity. After the first two steps of IMAIS identify the waste, the organization can treat the waste like any other problem and address it with CQI-10, DMAIC, or whatever corrective action process it normally uses. IMAIS consists of the steps in Table 3.1.

Isolate and Measure are therefore emphatically not identical to Six Sigma's Define and Measure. They correspond to and even precede Recognize in Six Sigma's Recognize, Define, Measure, Analyze, Improve, Control, Standardize, and Integrate (RDMAICSI) process. Harry and Schroeder (2000, 130) define the Recognize phase for the process level as "Recognize functional problems that link to operational issues" and Define as "Define the processes that contribute to the functional problems." Isolate and Measure are therefore the steps that recognize and hopefully quantify the waste, and therefore the need for subsequent DMAIC activities or their equivalent. Table 3.2 relates IMAIS to RDMAICSI.

Table 3.3, meanwhile, relates IMAIS to CQI-10. Isolate and Measure determine (1) whether waste exists and (2) where it is located, and they therefore precede as well as correspond to the first two steps of the CQI-10 process.

Isolate and Measure are meanwhile consistent and synergistic with the Supplier, Input, Process, Output, Customer (SIPOC) model and also the powerful material and energy balance analytical technique from the previous

Table 3.2 IMAIS and Six Sigma's RDMAICSI

IMAIS	*RDMAICSI*
Isolate the process with a control surface	(No corresponding step)
Measure the process's efficiency to identify gaps between actual and ideal or potential performance	Measure the process performance; *this comes before Recognize or Define* Recognize processes that require improvement
Assess the process for the root causes of the wastes in question	Define the specific issues (inefficiencies) Analyze the process; perform root cause analysis and other problem-solving activities to identify potential causes and solutions
Improve the process, and verify the effectiveness of the improvements	Improve
Standardize	Control Standardize Integrate

Table 3.3 IMAIS and AIAG's Effective Problem-Solving Guideline

IMAIS	*CQI-10 Effective Problem Solving*
Isolate and Measure	Problem Notification
	Problem Identification
(Not applicable to proactive improvement)	Containment
Assess	Failure Mode Analysis
	Root Cause Analysis
Improve	Choose and Implement Corrective Action
Standardize	Control and Standardize

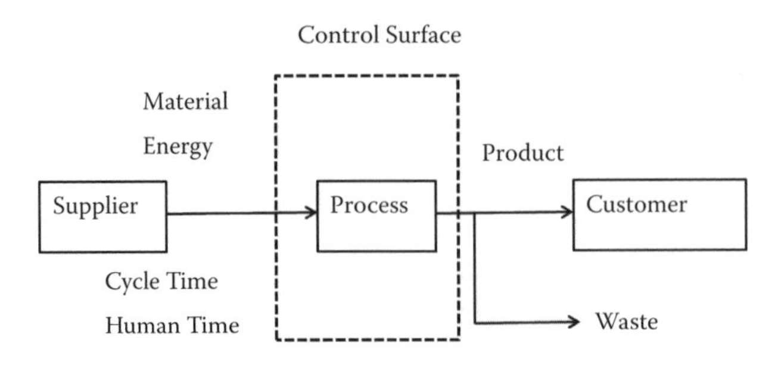

Figure 3.1 SIPOC model with control surface.

chapter. The first step is to isolate each process, which underscores the importance of the process perspective for analytical purposes.

Isolate

Isolate each process with a control surface for analytical purposes as shown in Figure 3.1. The control surface is the analytical boundary through which inputs and outputs flow and across which all inputs and outputs must balance. These include not only materials and energy but also the time of things and the time of people. IMAIS therefore recognizes cycle time and human time as resources like any other.

The previous chapter showed that the material and energy balance requires that inputs of material and energy equal outputs of products and waste. Metal cuttings (even recyclable ones); spent cutting fluids; expended chemicals, such as those used in semiconductor manufacture; and so on all constitute waste. The same applies to energy required to, for example, create the metal cuttings. Whether this waste is avoidable depends on the nature of the product and the process, but this analytical approach forces it all to become visible instead of being taken for granted. The following example explores a cadmium electroplating process.

Material Balance Example: Cadmium Electroplating

Figure 3.2 illustrates a cadmium electroplating process (Merit Partnership, 1996) to which a material balance might be applied. The item of most interest is the cadmium-contaminated wastewater. Performance of this material balance would be part of the Measure step of IMAIS, which is discussed in

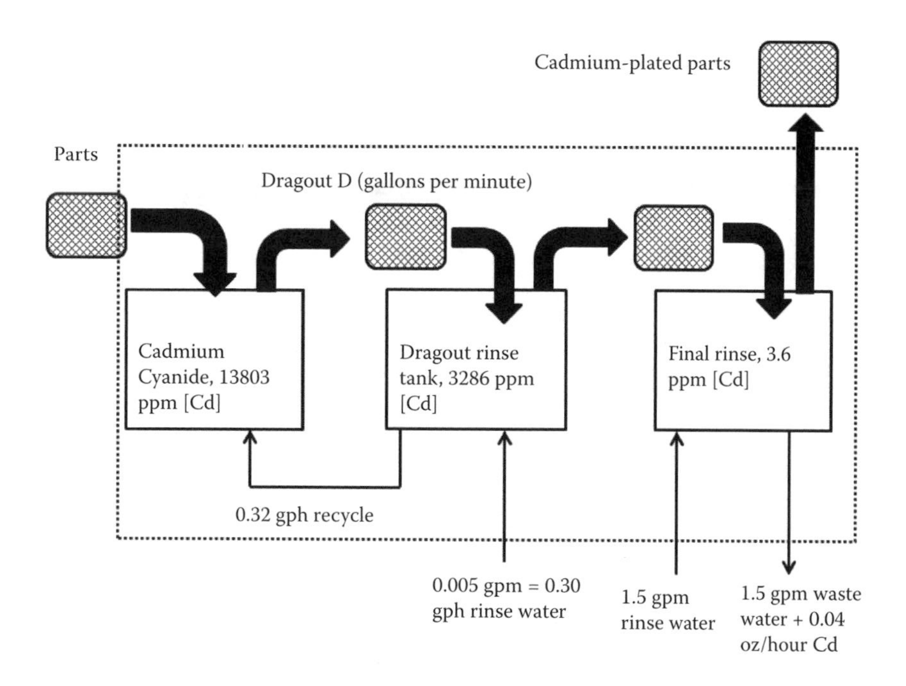

Figure 3.2 Isolation of a process with a control surface.

the Measure section. The first step (Figure 3.2) would be to isolate the entire process for analytical purposes with a control surface or control envelope.

"ppm [Cd]" in the context given means parts per million by weight as elemental cadmium, even though it is in chemical combination with cyanide.* Two material streams must balance across the entire system and across each tank: those of cadmium and those of water.

It is not possible to do a comprehensive material balance without knowledge of either the rate at which makeup cadmium cyanide is supplied to the plating tank or the rate at which cadmium metal is withdrawn by the finished parts. The control surface requires however that inputs match outputs in both quantity and kind. If for example the parts withdraw 1.96 ounces of cadmium per hour, and the wastewater as shown withdraws 0.04 ounces per hour, then 2 ounces of cadmium (as cadmium cyanide) must be supplied every hour if there is to be no accumulation or depletion in the system. Figure 3.3 shows that it is also possible to calculate the dragout rate D in gallons per minute, if it is assumed that D is constant from one tank to the next and also that the solution density is independent of the concentration. In this case, the control surface is placed around the final rinse tank.

* The usual context of [Cd] is moles of cadmium per liter of solution, or pound-moles of cadmium per gallon of solution, but calculation of the amount of cadmium in the waste stream shows the intention is ppm cadmium by weight.

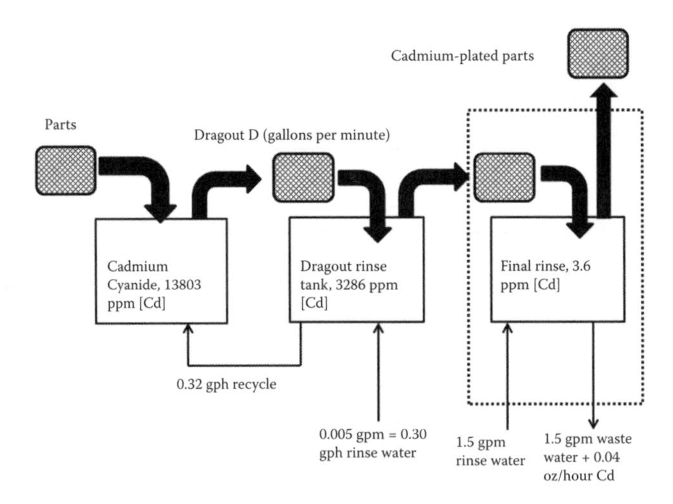

Figure 3.3 Dragout rate from a material balance.

Also assume that 0.00533 gallons per minute of rinse water (rounded down to 0.005 in the reference) are provided to equal the 0.32 gallons per hour of 3286 ppm solution that is recycled back to the plating tank; this must be true if solution neither accumulates nor depletes in the dragout rinse tank. The cadmium balance on the rinse tank is then as follows, with all figures in ounces of mass. The input to the rinse tank is the cadmium in the solution carried from the dragout rinse tank, and the outputs are the wastewater and the dilute solution carried from the rinse tank by the parts respectively, both at 3.6 ppm.

$$Cadmium : 0.003286 \times D\,\frac{gallons}{min}\,\frac{8.34\ pounds}{gallon}\,\frac{16\ ounces}{pound}$$

$$= \left(1.5\,\frac{gallons}{min}\,\frac{133.44\ oz}{gal} + D\,\frac{gallons}{min}\,\frac{133.44\ oz}{gallon}\right) \times 3.6 \times 10^{-6}$$

Ounces refer to mass and not volume in this context, noting the 16 oz/pound conversion. It is not really necessary to include the conversion factors because they all cancel out, but they are presented as a formality because the cadmium must balance by mass while the initial information is given by volume. The balance on water (assuming again that the solution's density does not change with cadmium cyanide concentration) can be performed on a volume basis. Solution for D as follows yields 0.001645 gallons per minute.

$$D = \frac{1.5 \times 3.6 \times 10^{-6}}{0.003286 - 3.6 \times 10^{-6}} = 0.001645$$

The next step is to do an overall material balance on the water per Figure 3.2, in which the control surface surrounds the entire process. The inputs are 0.00533 for the dragout rinse tank and 1.5 for the final rinse tank, while the outputs are 0.001645 for the parts that leave the rinse tank plus 1.5 for wastewater, with all quantities in gallons per minute. This shows that the system must somehow get rid of an additional 0.003685 gpm through evaporation or else level control of the rinse tank. In the latter case, the wastewater flow is slightly more than 1.5 gpm.

The reference gets 0.04 ounces per hour for cadmium in the wastewater stream, which is consistent with roundoff of the following results.

$$1.5\frac{gal}{min}\frac{133.44\ oz}{gal}\times 3.6\times 10^{-6}=0.000720\frac{oz}{min}=0.0432\frac{oz}{hour}$$

The problem is actually over-defined because a cadmium balance on the dragout rinse tank yields a slightly larger value for D. The input is the dragout from the plating tank, and the outputs are the recycled solution and the dragout from the dragout rinse tank respectively.

$$Cadium\text{: } 0.013803 \times D = (0.00533 + D) \times 0.003286$$

Solution of this equation yields D = 0.001665 gallons per minute. It is quite possible that the concentration in the rinse tank is not exactly 3.6 ppm; if it is 3.64 ppm, then a cadmium balance on the rinse tank yields D = 0.001663 gallons per minute as follows.

$$D=\frac{1.5\times 3.64\times 10^{-6}}{0.003286-3.64\times 10^{-6}}=0.001663$$

3.64 ppm rounds off to 3.6, and the analytical test for cadmium cyanide concentration might not even be capable of resolving the hundred millionths figure. There are in fact situations in which instruments and analytical tests return zero for extremely small concentrations, which is why statistical techniques for “nondetects” are of such great interest in pollution control activities.

The items of most practical interest here are the volume of wastewater (1.5 gallons per minute or slightly more) and the quantity of cadmium in it. The cost of the waste includes primarily the cost of treating the wastewater prior to disposal, and secondarily the value of the cadmium cyanide itself. The reference shows that an improvement of this process and a related chromate conversion process saved more in disposal costs ($2040 annually in rinse water, sewage, and wastewater filter media) than it did in cadmium cyanide and yellow chromate solution ($580/year). Chapter 14 will discuss the improvement further; installation of a counterflow rinse tank and a spray rinse prior to the dragout rinse halved the cadmium waste and, perhaps more importantly, reduced the volume of the waste stream by two-thirds.

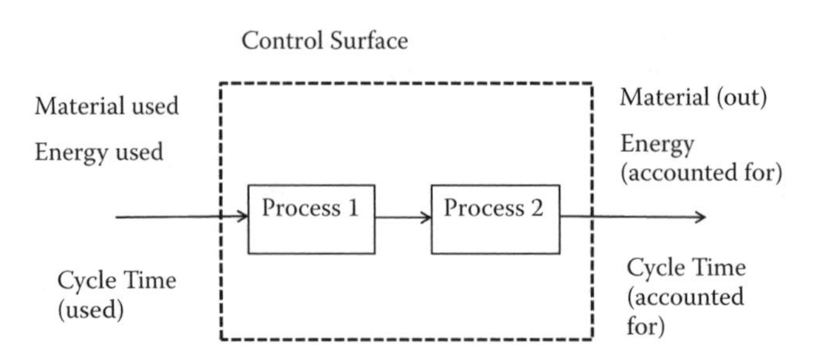

Figure 3.4 Discovery of hidden processes or handoffs by control surface approach.

A material and energy balance across the control surface therefore deals with two of the four recommended lean KPIs: waste of material and waste of energy. Cycle time accounting does the same thing for the time of things; it generates a detailed ledger for the resource of time. If the input consists of 100 minutes of cycle time, the time necessary for value-adding transformation, value-assisting handling and inspection, the necessary evil of transportation, and the complete waste of waiting must add up to 100 minutes.

Always remember that there is a huge opportunity for waste at process interfaces and process handoffs, especially if transportation is involved. These interfaces and handoffs should be assessed as separate processes, and the SIPOC model with control surfaces gives them no place to hide. Figure 3.4 applies this model to two (or by implication more) processes and shows that inputs of cycle time, energy, and material must balance outputs.

If, for example, the delivery of the product requires more cycle time than this assessment can explain, another process or handoff exists in the activity sequence that has not been taken into account. If more material goes in than comes out (as product or waste), something has not been taken into account, and the same goes for energy.

Isolate versus Supply Chain Perspective

"Isolate" sounds incompatible with a supply chain perspective. It is indeed vital to avoid suboptimization that creates more overall waste throughout the supply chain than it removes from any single element. Consider a Just-in-Time (JIT) process that relies on frequent less-than-truckload (LTL) shipments. Full loads would reduce waste of energy during transportation but increase cycle time (waste of the time of things) through the generation of inventory. The tradeoff can sometimes be resolved, however, through truck sharing, in which a truck may carry enough small shipments, possibly from different companies, to make up a full load.

"Isolate" therefore does not mean abandonment of the supply chain and value stream perspective. It is simply an analytical approach whose purpose is to force all wastes in a particular process to become visible. What, if anything, should be done about those wastes is part of the Assess step.

It is also important to define the process carefully. Energy is, of course, necessary to provide heating, ventilation, air conditioning, and lighting to an entire facility, but it is not possible to shut off these utilities around a single workstation that is not in operation. A large number of semiconductor processes may rely on the same cleanroom environment, but provision of that environment is not part of any individual process. Provision of suitable work conditions for the people and product must therefore be isolated as a separate process for analytical purposes.

Measure

Measure the performance of each process in terms of:

1. Waste of the time of things, as measured by manufacturing cycle efficiency (MCE).

   Smith (1998, 10) defines this as

$$\text{MCE} = \frac{\text{Value} - \text{adding time}}{\text{Total cycle time}}$$

   a. How much time does the work spend in the process in comparison to the value-adding time (specifically the time in which tools or the equivalent are transforming the work) that is nonnegotiable with the requirements of the process? What is the percent utilization of a capital asset for which value-adding work is available?
   b. Cycle time accounting requires categorization of all the time the work spends in the process, including time spent in transportation and time spent waiting for anything. This forces all time spent during handoffs and at process interfaces to become highly visible.
2. Waste of the time of people.

   a. Existing industrial engineering and human factors methods should cover this. Employees should also be intolerant of any job design that requires them to walk significant distances or requires excessive physical effort. Exertion does not equal value-adding work and is often a sign of waste effort.
3. Waste of materials.
   a. Material and energy balance.
   b. Bill of products or bill of outputs.
   c. "Dumpster diving" or visual observation of everything that is thrown away (subjective in comparison to the other techniques, which are quantitative and analytical).
4. Waste of energy.
   a. Material and energy balance.

The section on "Waste of Energy" stated that it is often possible to calculate energy requirements that are non-negotiable with the laws of nature. This enables a gap analysis that facilitates identification of waste.

Time necessary for transformation of the product, i.e., Masaaki Imai's value-adding "Bang!" is the non-negotiable cycle time, and it applies to services as well. If it takes a passenger airplane two hours to get from one city to another, this is the value-adding cycle time. Waiting on the runway is non-value-adding waste. Loading and unloading are handling, as are preventive maintenance and refueling.

This book described previously how the critical need to shoot more rapidly than an enemy drove motion-efficiency developments in armies, and the criticality of every second in an auto race has similarly driven the development of methods to get a car out of the pit and back onto the track as rapidly as possible. United Airlines has, in fact, recognized this (ABC *Good Morning America*, 2006) and sent employees to a school for NASCAR pit crews, whose job is to refuel a race car and change its tires in 13 seconds: "Our airplanes don't earn money while they're sitting on the ground," said Larry DeShon, senior vice president of airport operations for United. "They need to be in the air. So, if we can shave even four or five minutes off of every aircraft turn, we can fly well over a hundred more flights a day."

If the pit crew efficiency methods achieve this without the addition of additional personnel or aircraft—and the fact that the increase in capacity comes from rapid turnaround times instead of more airplanes—fuel and other consumables like oil and maintenance parts are the only marginal cost for the additional flights, which turns their marginal profit into pure profit.

Assess

The next step is to identify the root causes of the wastes that have been uncovered and determine whether they can be removed economically. Keep the supply chain concept in perspective: Assessment must include determination as to whether the proposed waste reduction will create greater wastes elsewhere. Examples are as follows.

Time of People versus Time of Things

Chapter 2 described how Ford subdivided tasks to eliminate non-value-adding motions, such as picking up and putting down tools. If one person places a fastener while another tightens it, the time necessary to pick up and put down the tool is eliminated. If the time necessary to place the fastener does not equal the time necessary to tighten it, though, the worker with the faster task will have to wait for the one with the slower task.

It might be desirable theoretically to somehow move the workload to make the task times equal, but this is not always practical. Meanwhile, Goldratt's Theory of Constraints (TOC) says that entirely equal task times might not be desirable because any variation whatsoever—in this case, in the time necessary to place the fastener and the time necessary to tighten it—will obviate most of the benefits. TOC also makes it clear that capacity imbalances are frequently unavoidable, and this concept applies to subdivided jobs, as well as pieces of equipment.

Note also that the inferior job design, in which each worker is kept "fully occupied" by picking up and putting down a tool for each fastener, wastes far more time of people than the subdivided job in which one worker might have to remain temporarily idle for each fastener. It also wastes far more cycle time. Manufacturing professionals should therefore not worry excessively if a more efficient job design leads to minor work imbalances. If the workers themselves feel that the arrangement is unfair, they can easily agree to switch tasks during the workday so each spends an equal amount of time at the least and most desirable operations.

Modern manufacturing involves far more assembly by machines than by people, though, which means machines and not people set the pace of the work. Japanese manufacturing firms are famous for reducing waste of the time of people by organizing workplaces so a worker can operate more than one machine simultaneously.

Energy versus Time of People and Time of Things

A truck requires less fuel to go 55 than 65 miles per hour, but this increases the cost of the driver by 18 percent and also wastes of the time of things. Operation of a container ship at half speed instead of full speed decreases the waste of energy (and fuel consumed per ton of cargo) enormously but increases waste of the time of the cargo and also the time of the crew. Consider the following example (Miller, 2009, website):

> On an early afternoon last month, the *Eugen Maersk* has left Rotterdam, the Netherlands, on the tail end of a journey from Shanghai. But the giant freighter is cruising at 10 knots, well shy of her 26-knot top speed. At about half speed, fuel consumption drops to 100–150 tons of fuel a day from 350 tons, saving as much as $5,000 an hour.

A 65 percent reduction in fuel consumption sounds very good until one notes the fact that Maersk would need 2.6 times as many 10-knot ships and crews to do the work of a 26-knot ship. The context of this situation was in fact an enormous plunge in global shipping volumes in 2009, which meant that Maersk could handle all its business at half speed with its existing capacity. This, on one hand, allowed Maersk to cut shipping costs, noting that $5,000 per hour in fuel savings more than offsets the crew's wages for the increased time at sea. The same applied to sailing around the Cape of Good Hope to avoid a $600,000 toll at the Suez Canal—a decision that itself involved the tradeoff of burning more fuel for the trip's increased distance. On the other hand, these decisions increased delivery lead times and inventory in the customers' supply chains: a situation that directly contravenes JIT manufacturing.

Note that the title of the Miller reference is "Shippers Taking it Slow in Bad Times." This underscores the fact that what might be an outstanding decision in bad times, i.e., when capacity far exceeds requirements, might be a catastrophic decision when excess capacity is not available or delivery lead times are critical. It is therefore always important to consider the "on the other hand" situation, which may change depending on the economic environment.

Materials versus Time of People

Ford and Crowther (1926, 97) described how the Ford Motor Company salvaged worn-out tools including drills, broaches, and reamers. Even tool handles were salvaged; a broken shovel handle might become handles for screwdrivers and chisels. Mop pails also were repaired and put back into service. Ford added the provision, "as long as it is profitable," which means there is a point at which the cost of labor to put a broken or worn-out item back into service exceeds the item's value. When a wooden item reached this point, Ford sent it to his wood distillation plant to convert it into wood chemicals and charcoal.

Improve

Implement the selected changes and verify their effectiveness. Do they remove the wastes that were previously identified and, if so, by how much? Are there any unforeseen and/or undesirable tradeoffs?

Standardize

The verified improvement is now the best known way to do the job in question, and it therefore becomes the standard. Update work instructions accordingly. Best practice deployment then requires identification of similar or related processes that might benefit from the same improvement or a similar one. This emphasizes the desirability of a keyword-searchable database of completed improvement projects. The Automotive Industry Action Group (AIAG, 2006, 137) describes a "Lessons Learned Database" that serves as a repository for this kind of knowledge.

"Read Across/Replicate Process" (AIAG, 2006, 172–173) is meanwhile the modern counterpart of Ford and Crowther's (1926, 85) statement that "An operation in our plant at Barcelona has to be carried through exactly as in Detroit—the benefit of our experience cannot be thrown away." The AIAG reference states,

> Reading across, or Replicating, the results of a problem solving effort is the last, but very important, step in the EPS [Effective Problem Solving] process. Replication simply means that all the

> results of a problem solving effort are replicated to benefit other products and parts of the organization.
>
> ...The primary purpose of Reading Across/ Replication is to drive the lessons learned and actions of one problem solving effort across to all other products and areas of an organization where applicable.

AIAG also provides an "Effective Problem Solving: Replicate Worksheet" for this purpose. Some revision may be necessary for application to Lean manufacturing, as "containment" applies primarily to quality deficiencies such as nonconforming parts. It is recommended that these worksheets be treated as quality records (and must therefore be controlled by a superior document, probably a second-tier procedure) and retained indefinitely.

Had the Ford Motor Company produced and retained such records during the 1910s and 1920s, the records would now furnish a treasure trove of case studies for Lean manufacturing practitioners. The limited number of examples cited by Ford and his contemporaries are enormously instructive, and one can only imagine how many thousands of similar examples have been lost.

Summary: IMAIS

Traditional improvement cycles begin with the assumption that a problem has been recognized, as might happen due to rework, scrap, or a customer complaint. Organizations that are somewhat more proactive empower a worker who sees a potential quality or safety problem to file a hiyari ("scare report") that initiates closed loop corrective action. Halpin (1966, 60–61) describes the *error cause removal* (ECR) program that is similar to the hiyari. The hiyari concept can be extended to quality system nonconformances, with a valid hiyari constituting the equivalent of an internal audit finding.

Even the hiyari and ECR require, however, the ability to recognize the potential deficiency for what it is. This is often but not universally possible for most wastes. Examples include visible material waste as might be discovered through observation or "dumpster diving," along with obvious waste motion such as walking to get parts. Inventory is symptomatic of the waste of the time of things, and once workers are taught to not take inventory for granted, they are likely to act on it. Arnold and Faurote (1915, 279–280) cite the presence of inventory where it did not belong, such as on a conveyor or work slide, as a visual signal of a stoppage or other problem. The only place inventory belongs is in the buffer of a capacity-constraining resource, as described by TOC.

However, an extension of Murphy's Law says that if it is possible for waste to go undetected, it will go undetected. IMAIS, in contrast to traditional improvement cycles and even the proactive hiyari and ECR, *sets out to look for trouble.* It does so analytically, with methods every bit as rigorous as debit and credit accounting. Material and energy wastes, the key considerations of ISO 14001 and ISO 50001, respectively, cannot hide from a comprehensive material and energy balance. Cycle time accounting forces all non-value-adding cycle time to become visible. Traditional industrial engineering and human factors engineering can meanwhile identify and eliminate waste of the time of people.

Although it is difficult to conceive how any form of waste in an operating manufacturing or service process can hide from IMAIS, *IMAIS cannot remove every waste that is designed into the product or process.* Lorenzen (1992) cites a Ricoh Copier study that showed that a design fix (or by implication improvement) offers a 100:1 payoff, a process fix a 10:1 payoff, and a manufacturing fix only a 1:1 improvement. The context was that of prevention or correction of quality problems, but the same lessons apply to Lean manufacturing. Waste that is designed into the product or process is often difficult to kaizen out, so the importance of Design for Manufacture (DFM) and Design for Assembly (DFA) cannot be overemphasized.

LMS:2012

II

Chapter 4

Lean Management System Requirements

Provision 4.1: General Requirements

The organization *shall* create, document, and continually improve a Lean management system (LMS). The system *shall* be able to identify and, where practical, suppress all wastes of (1) time, (2) materials, and (3) energy.

- The organization *shall* use a comprehensive quality management system (QMS) as a foundation for the LMS, although official registration to a QMS standard is optional.
- The LMS can and should be incorporated into an existing QMS or business management system (BMS) to avoid duplication of effort and conflicting directives.
- The LMS *shall* cover all processes that create or deliver the organization's products or services, as well as administrative and supplier processes upon which the organization depends. This includes outsourced processes as well as logistics.
- The LMS *shall* define criteria for the effectiveness of these processes and provide for their measurement and continuous improvement. These criteria *shall* include the following key performance indicators (KPIs), or alternatively other KPIs that encompass them:
  - Waste of the time of things: specifically the product or service and also capital assets for which paying work is available
  - Waste of the time of people

- Waste of materials (LMS/M)
- Waste of energy (LMS/E)

Explanation

The indicated KPIs encompass all forms of conceivable waste in an operating process, so omission of even one can allow waste to hide. If the organization already has KPIs with which it is comfortable, and they cover all the recommended ones (or can be revised to do so), this is equally satisfactory.

Planning and Assessment Questions

1. What QMS does the organization use?
2. What procedure does the organization use to identify its processes and the sequences and interactions (including handoffs from one to another) between them? The latter includes transportation, storage, and other non-value-adding activities.
3. How does the organization identify external processes (such as upstream supplier and downstream distributor processes) upon which it depends for its own operations?
4. How does the organization measure "Leanness" (or absence of waste)? What are the KPIs?
   a. What potential wastes can hide from the KPIs? The ideal answer will be the absence of an answer but "what" encourages people to look actively for examples instead of answering yes or no.
5. If the organization does not use the recommended KPIs, how do those it uses identify and quantify wastes of the time of things, time of people, materials, and energy?

Provision 4.2: Lean System Documentation

The organization *shall* document its LMS. Documentation of the LMS *shall not* conflict with that of the QMS, Environmental Management System (EMS), or Energy Management System (EnMS). The LMS can and should encompass both the EMS and EnMS.

Planning and Assessment Questions

1. What is the structure of the LMS? Four-tier hierarchies are typical but others may be used.
2. How does the LMS affect the planning, operation, and control of organizational processes? What procedures does it use to make sure that the appropriate Lean methods are applied to each process?
   a. Think of "processes" in terms of the Supplier, Input, Process, Output, and Customer (SIPOC) model, and remember that processes can extend outside the company's boundaries.
3. How is the LMS incorporated into the QMS?
   a. If the LMS is so incorporated, how are LMS-specific provisions differentiated from QMS provisions for which the organization is accountable to external auditors? The LMS provisions should be subject only to internal auditing, but a customer or registrar can audit any LMS provision that cannot be distinguished from a QMS provision.
4. How does the LMS ensure that changes in the first-tier Lean manual, a second-tier policy, a third-tier work instruction, or a fourth-tier quality record will not create a conflict with other documents?
   a. What procedure is used for cross-referencing superior, subordinate, and reference documents to make sure that their directives do not conflict? This also is an important audit question for the QMS itself.

Provision 4.2.1: General Documentation Requirements

The LMS documentation *shall* include:

- Statements of Lean policy and Lean objectives, which may be piggybacked onto or incorporated into the organization's quality policy and quality objectives
- A Lean manual, which can and should be incorporated into the quality manual
- Procedures and quality records to support the Lean policy and Lean objectives

Planning and Assessment Questions

1. How does management determine which policies, procedures, and quality records must be documented?
2. What are the procedures (generally second-tier documents) for identifying and, where practical, eliminating waste? The internal auditor should be able to ask this question successfully for all the Lean KPIs.
3. How does management review its documented procedures for:
   a. Adequacy (in terms of achieving the desired results)?
   b. Relevance to organizational performance?

Provision 4.2.2: Lean Manual

The organization *shall* create and maintain a Lean management manual. The Lean manual can and should be incorporated into the organization's quality manual to avoid duplication of effort and conflicting instructions. Its content *shall* include:

- The scope of the Lean management system.
- Procedures (generally second-tier documents) for the LMS, or references to them.
- Synergies and other interactions between the LMS procedures and between the LMS and the QMS.
- LMS provisions *shall not* conflict with the provisions of the QMS or other standards.
- LMS provisions should be identified as such to avoid confusion with QMS provisions for which the organization is responsible to external auditors.

Planning and Assessment Questions

1. Scope: How does the Lean management system extend through the relevant parts of the supply chain, beginning with suppliers and extending to distributors and customers?
2. What are the synergies between the LMS and QMS? That is, how do LMS provisions support basic quality objectives and vice versa?

Provision 4.2.3: Control and Retention of Documents and Records

The LMS defers to the prevailing QMS for specific document control and record retention requirements. LMS documents and records *shall* be treated exactly like QMS documents and records with regard to control and protection from loss or destruction.

Planning and Assessment Questions (Also Applicable to the QMS)

1. How long must each quality record be retained? (LMS-related records are treated as quality records.)
2. How are documents and records maintained and stored to protect them from loss or destruction?
   a. The auditor should be able to ask how the organization would retrieve the information if any selected records were destroyed.
   b. Are paper records stored in fireproof safes to avoid destruction by fire?
   c. Are copies made and then stored in a location separate from the originals?
   d. Are computer records stored in at least two separate places to avoid loss in the event of a fire or hard drive crash? Note that special fire safes are required to protect magnetic media such as CD-ROMs because temperatures insufficient to ignite paper are capable of destroying them.
3. What provisions are available for keyword searches of quality records, such as closed loop corrective actions, Quality Action Requests (QARs), Corrective Action Requests (CARs), proactive improvement actions, Failure Mode Effects Analysis (FMEAs), and so on?
   a. The record's existence shows only that a required action was completed. The concept of best practice deployment or the Lessons Learned database (AIAG, 2006, 137) requires, however, that related activities be able to benefit from the knowledge gained. See also Read Across/ Replicate Process (AIAG, 2006, 172–173).

Chapter 5

Organizational Responsibility

Provision 5.1: Organizational Commitment

- All levels of the organization *shall* show commitment to the Lean management system (LMS).
- Management *shall* earn the workforce's commitment by setting appropriate performance measurements and by participating actively and visibly in the Lean program.
- The organization *shall not* discharge any employee because of productivity improvements.
- The workforce or union (if there is one) *shall* show reciprocal commitment to the LMS.
- There *shall not* be any restrictive work rules that prevent any employee from performing any task that he or she can do and that needs to be done.

These provisions are unique to LMS:2012. No LMS can function properly if the workforce believes that productivity improvements will result in layoffs or if restrictive work rules prevent worker flexibility.

Planning and Assessment Questions

1. How does management communicate throughout the organization the importance of identifying and suppressing all forms of waste in the supply chain?

2. What reviews does management perform to support continuous improvement of Lean activities?
3. How do performance measurements support Lean manufacturing and Lean enterprise?
4. How does the cost accounting system promote dysfunctional operating decisions? (Recall that assessment questions should begin with a word like "how" as opposed to merely "does," which requires only a yes or no answer. The ideal response to "how does" in this case is a complete absence of identifiable answers, but the question also invites people to point out the answers if they exist.) The cost accounting system's measurements are suitable for tax and financial reporting standards. These measurements are often unsuitable for operational decision making, and they can have devastating consequences if they are misused for this purpose.
   a. When do dysfunctional metrics like "tool utilization" and overall equipment effectiveness (OEE) promote the production of unnecessary inventory?
   b. When do dysfunctional purchasing metrics encourage people to buy excessive quantities of raw materials? Such purchasing practices are generally not consistent with Just-in-Time (JIT) production, although there may be case-by-case exceptions.
   c. Where do metrics such as return on investment (ROI) have dysfunctional effects on operational decision making?
5. When do managers at all levels visit the value-adding workplace (gemba) as opposed to trying to run the operation from their offices? Tom Peters refers to this as "management by wandering around" (MBWA) and Masaaki Imai calls it gemba leadership. The Duke of Wellington referred to this practice as "taking trouble," or going to see a situation himself as opposed to reliance on word of mouth or other reports (Keegan, 1987, 149).
   a. How do managers and senior managers participate in training activities, even to the extent of training front-line workers personally?
6. How does management promote a business culture in which people focus on solving problems (or making improvements) instead of blaming other people?
7. How does the organization handle situations in which there is not enough work for all employees?

Provision 5.2: Customer Focus

LMS:2012 defers the issue of customer focus and customer requirements to the quality management system (QMS), noting, however, that these often include not only critical-to-quality functional characteristics but also Lean manufacturing considerations like cycle time and on-time delivery.

Planning and Assessment Questions (These Also Support the QMS)

1. How does the organization get the voice of the customer into the design and delivery of its outputs?
2. Who, including representatives from customers, suppliers, and other stakeholders, serves on cross-functional design teams for the product or service? Note the tie-in with Design for Manufacture and Design for Assembly. The factory also is a "customer" for the design, in terms of ease of manufacture.
   a. Where does the organization use Quality Function Deployment (QFD, also known as the House of Quality) and/or value analysis?
   b. If the organization uses Failure Mode Effects Analysis (FMEA), how does it make sure that failure severity ratings reflect the customer's requirements?
3. How does the organization identify the customer's Lean enterprise requirements such as short lead times, JIT delivery, and cost reduction?

Provision 5.3: Lean Management Policy

Top management is responsible for Lean management policy. This policy *shall* be consistent with the goal of eliminating all forms of waste as encompassed by the Lean key performance indicators (KPIs), and it *shall* be understood by all relevant supply chain participants. The Lean policy *shall* include environmental (LMS/M) and energy (LMS/E) considerations, noting that material and energy wastes are inconsistent with Lean manufacturing. Elimination of these wastes supports ISO 14001 and ISO 50001, respectively.

Sample Lean Policy (Not Part of Provision 5.3)

A single sentence such as the following can be easily memorized by all employees, and it meets the indicated requirements: "It is the policy of XYZ Corporation to identify all wastes of (1) time of things, (2) time of people, (3) materials, and (4) energy, and to remove these wastes from its supply chain wherever practical." The last part of the sentence recognizes that it may be more expensive to remove certain wastes than to tolerate them, and that there are sometimes tradeoffs between critical-to-Lean process characteristics.

Planning and Assessment Questions

1. What is the organization's Lean management policy?
   a. How does this policy address all forms of waste including time, materials, and energy?
   b. How does the Lean policy encourage and empower all members of the workforce to identify and eliminate waste?
   c. How does the Lean policy educate supply chain partners about the organization's Lean objectives?
2. How are Lean objectives established and reviewed?
3. How do the organization's performance measurements promote these objectives and measure their achievement?
4. How is the Lean policy communicated to all organizational participants, including upstream suppliers and downstream customers?
5. How is the Lean policy reviewed for suitability?

Provision 5.4: Planning for Lean Operation

Provision 5.4.1: Lean Objectives

Lean objectives *shall* be established and deployed at all relevant organizational processes and relevant parts of the *external supply chain*. These objectives shall be measurable and shall support the Lean policy.

Planning and Assessment Questions

1. What are the organization's Lean objectives?

2. How do the Lean objectives provide for the identification and elimination of all forms of waste? The auditor should keep the recommended Lean KPIs in mind with regard to "all forms of waste."
3. How do these objectives' performance measurements encourage their achievement?
   a. How do these performance measurements link to Lean-related customer satisfaction considerations like cycle time and cost reduction?
   b. How do the performance measurements tie in with the organization's business plan?
   c. What dysfunctional performance measurements encourage waste such as the production of unnecessary inventory? (The ideal answer consists of no identifiable answers, but asking "what" instead of "do" invites narrative responses instead of a yes/no response that may overlook a problem.)
4. How does the organization deploy its Lean requirements to upstream suppliers and downstream distribution networks?
   a. How does the organization ensure that these Lean requirements are synergistic with and not antagonistic to the interests of supply chain partners?

Provision 5.4.2: Lean System Planning

The LMS *shall* be planned to meet the requirements stated in LMS's general requirements in Chapter 4.

Planning and Assessment Questions

1. What analytical techniques does the organization use to plan and assess the LMS?
2. How do Lean planning and Lean auditing follow the process model?
   a. The auditor should think along Supplier, Input, Process, Output, Customer (SIPOC) lines; processes convert supplier inputs into outputs for customers. Suppliers and customers may be internal or external.
3. How do the analytical techniques and planning activities address process interfaces and handoffs between supply chain elements?
   a. Substantial waste can hide at these interfaces, with storage of inventory being an obvious example.

Provision 5.5: Supply Chain Responsibility

Provision 5.5.1: Responsibility

Top management *shall* ensure that responsibility and authority for the LMS are defined and communicated throughout the organization and relevant portions of its supply chain.

Planning and Assessment Questions

1. How does the organization promote cross-functional participation in improvement activities?
2. Who is empowered to initiate *closed loop proactive action* (CLPA)? What channels are available to do this?
3. Who is responsible for identification and elimination of waste for each process or operation? The internal auditor should be able to ask this question successfully for any process, and for all the recommended Lean KPIs.
4. Who is responsible for identification and elimination of waste in supplier processes and in interfaces between the organization and other elements of the supply chain?

Provision 5.5.2: Management Representative

A specific member of management *shall* have responsibility and authority for implementing and maintaining the LMS. The management representative *shall* keep top management informed of the performance of the LMS. The management representative can be the same person who is responsible for the QMS to improve synergies while reducing the opportunity for system conflicts and duplication of effort.

Planning and Assessment Questions

1. Who is the LMS management representative?
2. How does he or she keep top management up-to-date on the performance of the LMS?

Provision 5.5.3: Supply Chain Communication

There *shall* be appropriate communication channels throughout the organization and relevant portions of its supply chain to support utilization and improvement of the LMS.

Planning and Assessment Questions

1. What communication channels support of the LMS? These can include but are not limited to:
   a. Management-led training and communication meetings.
   b. Presentations by work teams to management and/or other work teams.
   c. Poster boards (e.g., describing improvement projects).
   d. Intranets and internal websites, including keyword-searchable data bases of completed projects (supports best practice deployment). These can include AIAG's (2006, 137) Lessons Learned database.
   e. E-mail.
   f. Employee and work-team initiation of improvement projects (e.g., the hiyari or scare report).
   g. Electronic and other communication links between supply chain partners.
2. In what ways does the organization make itself porous (Tom Peters's terminology) to supply chain partners as well as its own personnel or break down organizational barriers as recommended by W. Edwards Deming and Henry Ford? What communication channels are available for customers and suppliers, both internal and external?
   a. The idea is that barriers such as functional silos, bureaucratic impediments, and so on should not impede internal or external communications.
   b. Note that this consideration also supports ISO 9001:2008 elements 5.2 (Customer Focus) and 7.4 (Purchasing).
3. What communications are retained as quality records?
   a. Records of training and communication meetings should be treated as such.

Provision 5.6: Lean System Review

Provision 5.6.1: General Requirements

Top management *shall* perform scheduled reviews of the LMS to ensure its effectiveness and make improvements. Review policy *shall* conform to that of the QMS. Records of management reviews *shall* be retained according to the organization's record retention policy.

Planning and Assessment Questions

1. What is the schedule for reviews of the LMS?
2. When was the last review performed and when will the next one take place?
3. What items does top management review, and why?
4. Munro (2004, 96–97) cites Safety, Quality, Delivery, Cost, Morale, and Environment (SQDCME) and Quality, Cost, Delivery plus Environment (QCD+E) as QMS review items. This book recommends wastes of the time of things, time of people, materials, and energy as KPIs, but others may be more suitable for individual organizations.

Provision 5.6.2: Review Input

Inputs to management review *shall* include but are not limited to the following, as applicable:

- Audits of the LMS
- Customer feedback as to the organization's ability to meet Lean-related requirements such as low costs and cycle time (if these are not already covered by the QMS)
- Production control feedback
- Process efficiency and effectiveness, including relevant upstream and downstream supply chain processes
- Status of proactive improvement projects and open items from previous management reviews

Planning and Assessment Questions

1. Customer feedback: Are customers with JIT production control systems satisfied with the organization's ability to provide JIT delivery and short lead times for order turnaround?
2. Production control feedback: The organization's production control system is a customer of upstream suppliers. Do these suppliers provide satisfactory JIT delivery?
3. Project status: What is the procedure for ensuring completion and closure of open items?
4. Munro (2004, 97–98) recommends a *QMS failure mode effects analysis* (QMS-FMEA), which sounds similar in principle to a process FMEA. Has the organization performed an LMS-FMEA to identify, for example, modes by which the LMS might fail to identify and address waste (muda)?
   a. This leads to another question: When employees discover wastes that were not visible to the LMS's key performance indicators, how does the organization modify the system to keep similar wastes from hiding in the future?

Provision 5.6.3: Review Output

The output of the management review *shall* include but is not limited to:

- Improvement of the LMS
- Responsiveness to customer Lean requirements, such as but not limited to JIT delivery, shorter lead times for order turnaround, and design for manufacturability in the customer's processes

Chapter 6

Lean System Infrastructure and Resources

Provision 6.1: Resource Availability

The organization *shall* identify and provide the resources necessary to implement, maintain, and improve the Lean management system (LMS).

Planning and Assessment Question

1. How does provision of resources, including time and attention, underscore management's commitment to the LMS? When does withdrawal of resources convey the impression that management doesn't think the LMS is all that important?

Provision 6.2: Workforce Training and Empowerment

All levels of the workforce *shall* be trained and empowered to identify and eliminate waste from the activities for which they are responsible. Training and education records *shall* be maintained as required by the organization's quality management system (QMS).

Planning and Assessment Questions

1. What are each job title's job description and the requisite Lean enterprise skills that go with it?
   a. How does the organization determine the Lean skills and competencies that are necessary in each position?
   b. Where does the organization's *training needs assessment* (TNA) system address Lean competencies?
2. How does the organization train and empower all its members to identify and eliminate waste?
3. How does the organization assess the effectiveness of Lean training and education?
4. What cross-functional team structures does the organization use to empower personnel to identify and eliminate waste? (Examples might include self-directed work teams, cross-functional project teams, and teams that include customer and supplier representatives.)
5. How does the organization make sure that *task significance* plays a role in each worker's performance?
   a. How does the organization assess or measure workers' awareness of the significance of their jobs' outputs to internal or external customers?
6. How does the organizational culture support a zero-tolerance attitude toward all forms of waste, including seemingly minor forms that people can largely ignore or work around?
   a. What mechanisms are in place to allow people to act on waste that they identify? (Chapter 8's section, "Proactive Action," covers procedures like the hiyari or "scare report" that allows workers to initiate corrective and preventive actions.)

Provision 6.3: Facilities, Layout, and Supporting Services

The organization *shall* provide facilities, layout, and supporting services (including information systems) conducive to Lean operation and waste reduction.

Planning and Assessment Questions

1. How do the facilities (building and utilities) and supporting services like transportation, communication, and information systems support Lean production?
2. What computerized pull production control system is used?
3. How does the factory layout reduce (or aggravate) non-value-adding transportation?
   a. Where do work cells, work slides, and conveyors replace carts, trucks, forklifts, and other methods of batch transportation?
   b. Health care: How does the hospital layout minimize the walking that nurses and other health care professionals must do? (The Henry and Clara Ford Hospital was designed to minimize unnecessary walking.)
   c. Where are supplies, hand-washing stations, and other resources positioned to make them easily accessible to health care workers?

Provision 6.4: Work Environment, Ergonomics, and Motion Efficiency

The organization *shall* provide a work environment, including ergonomics and motion efficiency considerations, that is conducive to Lean operation. The workforce itself *shall* be treated as an *internal customer* for the work environment.

Planning and Assessment Questions

1. How does the organization ensure ergonomic design of all physical tasks?
   a. Which jobs require a worker to routinely take more than one step in any direction or bend over?
   b. Which jobs require a worker to routinely apply large physical forces or "elbow grease"?
   c. Which jobs require a worker to routinely lift large loads?
2. How does the organization identify and implement appropriate workplace lighting?
3. How do heating, ventilation, and air conditioning promote a comfortable work environment?

4. How does the workplace mitigate excessive noise that, even if acceptable from an industrial hygiene perspective with ear protection, interferes with worker communication and audible process controls on the shop floor?
5. How does the organization ensure workplace cleanliness? Is a program such as 5S-CANDO (Clearing up, Arranging, Neatness, Discipline, Ongoing improvement) in place?
   a. When do workers waste time searching for tools or supplies?
   b. Where are waste containers accessible in the workplace? Where are waste containers *not* readily accessible? Omnipresent waste containers kept Henry Ford's factories meticulously clean, and Disney theme parks later adopted the same practice.

Chapter 7

Product or Service Realization

Provision 7.1: Planning

The organization *shall* plan, design, and develop processes so as to remove waste of time, waste of materials, and waste of energy from the product or service.

Planning and Assessment Questions

1. What tools and analytical techniques identify waste? Examples might include:
   a. Value stream mapping, with classification of activities as value-adding, value-enabling (such as preventive maintenance, process control, error-proofing), or non-value-adding.
   b. Material and energy balance.
   c. The Bill of Outputs (Levinson, 2005) is similar in principle to the material and energy balance, but it is designed for discrete product manufacture. It forces all outputs, including machining wastes, cutting fluids, depleted chemicals, and so on to become visible.
   d. Cycle time accounting.
   e. Spaghetti diagrams.
   f. Traditional industrial engineering and human factors techniques that identify waste of the time of people and/or poor ergonomics.

Provision 7.2: Customer Lean Operation Requirements

The organization *shall* identify, review, and implement customer Lean operation requirements such as but not limited to Just-in-Time (JIT) delivery and small-lot delivery.

Planning and Assessment Questions

1. How does the organization meet the customer's JIT delivery requirements?
   a. Is the organization capable of making to order instead of making to forecast? This can be answered yes or no; if goods are produced to forecast, or if there is an inventory of unsold finished goods, the organization is obviously not producing to order.
   b. Are order fulfillment lead times short enough to meet customer JIT requirements? The presence of inventory beyond customer requirements or, on the other hand, late deliveries and ensuing shortages for the customer indicate an answer of "No."
2. How does the organization identify and implement customer requirements that relate to error proofing? (This also relates to the customer's quality requirements.)
   a. Does the product have slots, keys, alignment markings, or similar error-proofing features when required by the customer's processes?
3. What communication methods does the organization use to identify customer needs and requirements? How does the organization make itself porous (Tom Peters's term) to its customers or eliminate communication barriers with its customers? (This question also applies to supply chain management, covered later in this chapter.)

Provision 7.3: Product, Process, and Service Design for Lean

The organization *shall* design and develop products to remove waste from manufacturing, transportation, and delivery. Services *shall* be designed to reduce or eliminate waste. LMS:2012 otherwise defers to the quality management system (QMS) for design and development-related activities.

Planning and Assessment Questions

1. How does the organization incorporate Design for Manufacture (DFM) and Design for Assembly (DFA) into its product design strategy?
   a. Note that the manufacturing department is an internal customer for the design.
2. How are products designed to reduce or eliminate the need for material removal processes?
3. How does the manner in which the product is packaged or shipped conserve (or waste) energy required for transportation?
4. How are products, processes, or services designed to minimize energy requirements?
   a. What techniques are available to reduce or eliminate the need for deburring, heat treatment, and similar operations?
5. Where does the DFM/DFA strategy use the following techniques, if appropriate? (See Cubberly and Bakerjian, 1989, for details on these techniques).
   a. Modular designs.
   b. Standard parts and components.
   c. Incorporation of error-proofing features such as keys, slots, and alignment markings into part design.
   d. Minimization of the use of fasteners (noting that the cost of driving a screw can be five or ten times the cost of the screw itself).
   e. Avoidance of flexible components like wires and wire harnesses.
   f. Use of *compliant* parts that do not require excessive force for assembly.
   g. Use of *group technology,* which groups parts into *production families* that use the same tooling sequence. This means that a work cell with a given tooling set can make any member of a production family.
6. Where does the organization use Taguchi's Robust Design and/or Design for Six Sigma (DFSS) (if applicable)?
7. How are tools and equipment selected to reduce overall energy consumption?

Provision 7.4: Purchasing: Lean Supply Chain Practices

Purchasing practices *shall* follow Lean principles.

Planning and Assessment Questions

1. What dysfunctional performance measurements encourage the purchase of excessive quantities of material that are not necessary for immediate use?
2. How does the purchasing department work with suppliers to achieve JIT delivery?

Provision 7.4.1: Deployment of Lean Requirements to Suppliers

The organization *shall* ensure that suppliers meet Lean manufacturing requirements such as JIT delivery, short lead times, and cost reductions. The organization *shall*, where appropriate, engage in supplier development activities to help suppliers meet these requirements. This section defers to the QMS for all other purchasing-related considerations.

Planning and Assessment Questions

1. How does purchasing policy implement *zero base pricing* (ZBP)?
   a. Schonberger (1986, 157) explains Polaroid's ZBP approach as follows: "Buyers are not to accept cost increases as justification for a price rise. Instead, at that point, Polaroid people visit the supplier's plant and offer tips on how to contain costs."
   b. How does the organization practice *supplier development*? Teaching suppliers how to use Lean methodology to reduce costs and lead times supports ZBP by helping suppliers keep their costs down. It may also be necessary to teach suppliers how to become effective JIT supply chain partners.
2. How does the purchasing function measure suppliers' Lean performance?
   a. Which suppliers deliver JIT-sized quantities, and which deliver large lots that must be held as inventory?
   b. Which subcontractors have unacceptably long lead times?
3. What Lean methods does the purchasing function use to communicate with suppliers?
   a. How does the organization tie its production control system to those of its suppliers electronically to facilitate JIT ordering?
   b. What forms of kanban (electronic or otherwise) does the organization use to replenish vendor-supplied parts and materials?
   c. If Drum-Buffer-Rope (DBR) is in use, do customer constraints "beat the drum" to set the pace for supplier constraints?

Provision 7.5: Lean Production and Service

The organization *shall* plan and carry out production and service operations to minimize all wastes as defined by the recommended Lean key performance indicators (KPIs).

Planning and Assessment Questions: TPS Seven Wastes

Per Taiichi Ohno, 1988, 19–20.

1. How does the organization reduce the waste of overproduction?
   a. What form of pull production control does the organization use to prevent the generation of unnecessary inventory? Methods include but are not limited to kanban production control and synchronous flow manufacturing's DBR production control.
   b. Where does heijunka (production load smoothing or leveling) apply to the organization's activities?
   c. How does the organization coordinate its production control system with customers and suppliers? The idea is that a good pull production control system can operate across a good part of the entire supply chain, with downstream customers pulling work from upstream suppliers.
      i. If the organization is using DBR, for example, can it tie its DBR system to those of its customers and suppliers?
      ii. Can the organization use "electronic kanban" to pull shipments from its suppliers? Can it respond to "electronic kanban" from downstream customers?
2. How does the organization reduce the waste of waiting and time in queue?
   a. How does the organization reduce reliance on batch production while achieving single-unit flow production?
   b. What methods are used to reduce setup times? Consider single-minute exchange of die (SMED) or related methods.
   c. How is variation in processing times and material transfer times (which increases the need for protective inventory buffers) suppressed?
   d. If there is a constraint or capacity-constraining resource (CCR) as defined by the Theory of Constraints (TOC) how does the organization ensure that it is not idled by lack of work (starving the constraint), lack of personnel, or unplanned downtime?

   e. What preventive maintenance techniques does the organization use to prevent unplanned downtime?
      i. What computerized maintenance management system (CMMS) supports the preventive maintenance system?
   f. What procedures ensure that parts necessary for maintenance will be available when they are needed?
      i. Waiting for parts often contributes to unplanned downtime.
   g. What is the policy toward so-called minor stoppages that result in very little downtime?
      i. National Semiconductor's policy is that there is no such thing as a "minor" stoppage (Gardner and Nappi, 2001). Such stoppages may be symptomatic of more serious problems, and they always waste time.
      ii. It is worth restating Halpin's (1966, 60–61) description of waste, muda, or friction: "They turned out to be the little things that get under a worker's skin but are never quite important enough to make him come to management for a change." Because workers can clear so-called minor stoppages themselves, they often learn to live with a recurring problem, and it never comes to the attention of management.
3. What methods reduce the time work spends waiting to be processed? Smith (1998, 10) states that manufacturing cycle efficiency, the proportion of value-adding time to total time, *is often less than 1 percent*. This is consistent with the analogy of a golf game (Standard and Davis, 1999, 61) in which the golf club (tool) acts on the ball for about 1.8 seconds in a four-hour game.
4. How does the organization reduce the waste of transportation? (The following list includes possible ways "how" might be answered.)
   a. Where does the organization use spaghetti diagrams or a similar technique to identify the waste of interdepartmental transportation?
   b. Where are conveyors or work slides used (where appropriate) to allow the easy transfer of parts from one operation to another?
   c. Where does the organization use work cells or unitary machines to reduce transportation as well as handling?
5. How are non-value-adding activities identified and removed from the process?
   a. Where does the organization use process flowcharting and/or value stream mapping to identify non-value-adding activities?
6. How is the waste of inventory reduced?

   a. How does the organization identify wasteful inventory in its supply chain, such as float or materials in transit?
      i. A "JIT warehouse" (which stores inventory for delivery in JIT-sized lots) is an oxymoron; it simply moves the inventory off site.
      ii. A container ship is a floating warehouse.
7. How is waste motion identified and reduced?
   a. Are jobs videotaped and reviewed by the workers who perform them? The videos' purpose is to assess not the workers but rather the way their jobs are designed. People often become accustomed to waste motion after a while and "learn to live with it," but it often becomes obvious when they watch themselves doing the job.
   b. How does the organization apply ergonomics and human factors to job design?
8. How does the organization reduce the waste of poor quality?
   a. The QMS should address a good part of this item.

Planning and Assessment Questions: Lean Key Performance Indicators

1. How does the organization identify, measure, and reduce waste of the time of things (cycle time)?
   a. How does the organization distinguish between value-adding cycle time (transformation processes, or actual delivery of a service); necessary evils like transportation, handling, and setup; and non-value-adding waste such as waiting?
   b. What is the manufacturing cycle efficiency (MCE) for any given process? The internal auditor should be able to obtain an answer for any process.
2. How does the organization identify, measure, and reduce waste of the time of people?
   a. What jobs require people to walk to get or move parts or wait at supply cribs or stockrooms?
   b. How does the organization identify and eliminate waste motion, ergonomically inefficient motions, and so on?
3. How does the organization identify, measure, and reduce waste of materials?
   a. What is the material and energy balance for this process? (The auditor should be able to obtain this for any process.)

b. What happens to materials that are not incorporated into the product? Ideal answers include reuse, recycle, or convert into saleable by-products. "Pay for disposal," e.g., by scrubbing pollutants out of a waste stream or by shipment to a landfill, indicates a need to look for a better alternative.
c. How much of the active chemical in, for example, an etching or plating solution is actually used on the product or incorporated into the product?
d. What methods does the organization use to avoid the production of material waste in the first place? This applies to wastes that are not environmental aspects as well as those that are.
e. How are materials such as chemicals and cutting fluids recycled or reused until they are no longer serviceable?
f. In what ways are waste materials sold or otherwise turned to alternative uses? Ford, for example, processed the slag from his blast furnaces into cement and paving materials while disposing of hardwood chips from lumber operations as Kingsford charcoal briquettes. *Ford News* (1923, No. 15, p. 6) adds that the company held a contest in which employees suggested uses for the waste ends of wooden spokes. The first prize was awarded for wood distillation, but other recommendations included wooden flooring and wooden mats. The committee judged both of these applications to be practical and said the wooden mats would be useful in the factory. Two other employees were given honorable mentions for recommending the use of shorter rough stock to avoid production of the waste in the first place and also to reduce its transportation costs and use of space in the kiln used for drying the wood. This single article, in fact, gives enormous insight into the thought process that prevailed through the entire Ford enterprise during the early 1920s.
g. In what applications do the organization and its supply chain neighbors use returnable or reusable packaging? The packaging should be broken down and sent back for another load or else relabeled and used to package the organization's own products.
h. How do painting and coating operations minimize waste of material? Shigeo Shingo pointed out the importance of painting the product and not the air. What he meant was that the spray paint that missed the product was wasted material and also an environmental problem (Robinson, 1990, 101–102).

   i. How are machining processes designed to minimize waste? Robinson (1990, 103) suggests cutting large holes instead of drilling them. A drill must expend energy to reduce the hole's entire cross section to shavings, whereas a cutter does this only to a thin cylindrical section.

4. How does the organization identify, measure, and reduce waste of energy?
   a. The question about material and energy balances applies here as well.
   b. How does the process recover or reuse mechanical and other forms of energy that would otherwise go to waste?
   c. How does the organization bypass the unavoidable inefficiency of thermal power cycles? Fuel cells, which convert chemical energy directly into electrical power, bypass the efficiency limitations of power generation cycles that transfer heat from hot reservoirs to cold reservoirs.

Provision 7.5.1: Lean Process Control

The organization *shall* use Lean process control techniques where appropriate. This section otherwise defers to the QMS for all activities related to production and service provision.

Planning and Assessment Questions (Many of These Also Support the QMS)

1. Where are *visual controls* such as andon lights, kanban boards, and kanban squares used?
   a. 5S-CANDO (arranging aspect): Where is there a clearly marked location for any important tool or other item the internal auditor may select?
   b. Where do andon lights indicate the status of production equipment?
   c. Error proofing or poka-yoke: How do color codes or other markings help ensure proper product assembly and equipment operation?
   d. How do statistical process control (SPC) charts make the status of the process (in or out of control) clearly visible to the workers?
   e. How are safety-related markings made clearly visible to all personnel? (LMS/S: supports safety and industrial hygiene.)

2. What error-proofing techniques are used?
   a. How do part and equipment designs make it impossible to assemble parts the wrong way or from the wrong direction?
   b. How do self-check systems, snap gages, or go/no-go gages prevent the transfer of nonconforming parts?
   c. What autonomation (jidoka) techniques allow tooling to detect and react to nonconforming conditions?
3. How do jobs incorporate the Ford safety principle “Can't rather than don't,” which is a form of error proofing (LMS/S: supports safety and industrial hygiene)? This means the job is designed so the worker cannot have an accident, as opposed to having to remember “Don't do so-and-so.” It is particularly applicable to health care environments, in which the medical worker cannot forget to practice good hand hygiene, give a patient the wrong medication, or connect a feeding tube to an intravenous line.

Provision 7.6: Control of Gages and Instruments

The LMS defers to the prevailing QMS on this item.

Provision 7.7: Supply Chain Management

The organization shall assess and continuously improve the Lean performance of the supply chain in which it is a participant.

Provision 7.7.1: Customer–Supplier Relations

The organization *shall* assess and continuously improve the working relationship between its external suppliers and customers.

Planning and Assessment Questions

1. What level of trust exists between the organization and its suppliers and customers?
   a. Walker (2001a) warns that lack of trust is the chief barrier to successful supply chain management. Trust and good communications are prerequisites for successful supply chain operation.

   b. When does the organization (or its customers) feel a need to place orders before they are really needed, for fear that the supplier will not deliver in time?
   c. Which suppliers are afraid that customers will place orders "just in case" and then cancel if the parts or services are not needed?
   d. What techniques does the organization use to avoid suboptimization, in which each supply chain participant tries to maximize its own profit at the expense of the others?
   e. How does the organization avoid dysfunctional behavior because of internal transfer pricing conflicts between internal customers and suppliers? This is the internal analogue of suboptimization.
2. How does the organization communicate with its supply chain partners?
   a. What long feedback loops undermine timely reaction to changing market conditions?
3. What procedures does the organization use to make it porous (Tom Peters's term) to its customers and suppliers?
   a. What departmental barriers must external customers and suppliers negotiate when dealing with the organization? The ideal answer is that the customer or supplier can go directly to the person who can resolve their questions or problems.
   b. When does the organization use customer contact teams (CCTs) or cross-functional teams that may visit a customer's place of business to see how the organization's product or service is used? When does the organization invite its suppliers to send similar teams to its own place of business?

Provision 7.7.2: Transportation

The organization *shall* assess and continuously improve the transportation systems through which it receives resources and delivers its product or service.

Planning and Assessment Questions

1. How does the organization evaluate the transportation systems that it uses?
2. Where are freight management systems (FMSs) and third-party logistics (3PL) systems used to reduce transportation costs and/or delivery times?
3. When does the organization use truck sharing, "milk runs," or other appropriate techniques to handle less-than-truckload (LTL) applications?

4. How does the company factor transportation considerations into its purchasing decisions?
   a. Where do "world sourcing" purchasing practices, in which companies shop the entire world for the lowest price, undermine the supply chain's Leanness? Note that container ships are essentially floating warehouses that add weeks to cycle times while forcing suppliers to ship large batches.

Chapter 8

Measurement and Continuous Improvement

Provision 8.1: Measurement and Analysis for Continuous Improvement

The organization *shall* plan and implement measurement systems necessary to ensure conformity of the Lean management system (LMS) and drive its continuous improvement. These systems *shall* be capable of assessing the system's ability to suppress waste of time, materials, and energy as defined by the recommended key performance indicators (KPIs).

Provision 8.2: Monitoring and Audit

Provision 8.2.1: Satisfaction of Customer Lean Requirements

The organization *shall* measure its satisfaction of customer Lean requirements such as Just-in-Time (JIT) delivery, short lead times for order fulfillment, and cost reduction.

Provision 8.2.2: Internal Audit

The organization *shall* perform scheduled internal audits to ensure that the LMS's structure and execution meets the requirements of LMS:2012 and also to identify opportunities for improvement. The audit *shall* specifically assess corrective action for nonconformances that were found in previous audits. These audits can and should be piggybacked onto quality management system (QMS) audits to avoid duplication of effort.

The organization should empower personnel to identify LMS nonconformances and initiate closed loop corrective action for them at any time. This may be achieved through the hiyari ("scare report") or any other effective method.

Planning and Assessment Questions

1. How does the organization ensure that nonconformance root causes that may cross departmental boundaries are identified and resolved? (A nonconformance in one area is often symptomatic of an organization-wide problem, and correction of that problem is an example of best practice deployment.)
2. How does the organization empower employees or work teams to submit an LMS (or, for that matter, a QMS) hiyari (scare report) at any time that carries the same weight as an audit finding? This is *not* required by ISO 9001:2008, but it is highly recommended. It effectively makes everyone an auditor and promotes continuous checks of the system.
   a. How does the organization empower employees or work teams to file a safety hiyari or its equivalent? (LMS/S: supports safety and industrial hygiene.)
3. How are internal and second-party auditors trained and qualified?
   a. As there are currently no registrars that audit to any kind of Lean manufacturing standard, the organization must decide for itself how to train and qualify its internal auditors. The same goes for second-party auditors if it wishes to audit its suppliers for Lean manufacturing practices. The American Society for Quality offers a certification in quality auditing, and the same basic principles apply to auditing in general. The Society of Manufacturing Engineers began to offer certifications in Lean manufacturing in 2006.

Provision 8.2.3: Measurement and Monitoring of Process or Service

Processes and services *shall* be measured and monitored for waste of time, materials, and energy as encompassed by the recommended KPIs, and corrective or proactive action *shall* be taken to remove such wastes where practical. ("Where practical" means the improvement is economically feasible and will not create a greater waste elsewhere.)

Provision 8.2 otherwise defers to the prevailing QMS because ensurance of conformance to specifications is an aspect of quality management.

Planning and Assessment Questions

1. How does the organization measure and monitor the product or service to minimize the cost of information collection?
2. What Lean-related information is collected for the product or service?

Provision 8.3: Containment of Nonconforming Product or Service

LMS:2012 defers to the prevailing QMS on control of nonconformity.

Provision 8.4: Data Analysis

The organization *shall* collect and analyze data to assess the effectiveness of the LMS and identify opportunities for its improvement.

Planning and Assessment Questions

1. How does the organization assess fulfillment of customer requirements such as JIT delivery, short lead times, and cost reduction?
2. How does the organization measure process performance?
   a. How do measurements quantify value-adding versus non-value-adding work?
   b. Where does the organization use material and energy balances (as in chemical process industries) or a bill of outputs to force all forms of material and energy wastes to become visible?
   c. How does the organization quantify waste of the time of product (or service) and waste of the time of people?
3. How does the organization assess supplier performance?
   a. Do metrics include JIT delivery, short lead times, and cost reduction?

Provision 8.5: System, Process, and Service Improvement

Provision 8.5.1: Continuous Improvement

The organization *shall* continuously improve the effectiveness of the LMS through all appropriate methods.

Provision 8.5.2: Proactive Action

The organization *shall* establish a documented procedure to support proactive suppression of wastes of time, materials, and energy. This procedure *shall* include a *closed loop proactive action* (CLPA) procedure that follows the traditional Plan, Do, Check, Act (PDCA) improvement cycle such as but not limited to:

- Ford Motor Company's Team-Oriented Problem Solving, Eight Disciplines (TOPS-8D) procedure. TOPS-8D can be used as it is, minus the containment of poor quality step, for CLPA projects. (Since proactive action occurs in the absence of poor quality, 8D's containment step is not applicable.) AIAG's CQI-10 is similar and also more recent.
- Six Sigma's Define, Measure, Analyze, Improve, Control (DMAIC).
- Lean-specific Isolate, Measure, Assess, Improve, Standardize (IMAIS).
- Other variants of PDCA.

The CLPA can and should piggyback onto the closed loop corrective action (CLCA) procedure that is already required by the major QMS standards.

Planning and Assessment Questions

1. How are CLPA projects initiated? How can personnel or work teams initiate a CLPA when they identify waste?
   a. Halpin (1966, 60–61) discusses the error cause removal (ECR) system, which is similar in concept to the hiyari.
2. How does the CLPA require:
   a. Definition and identification of the project team (e.g., project champion with the resources and authority necessary to implement the solution, subject matter experts, facilitators, and team leaders)?
   b. Definition of the problem or opportunity to be addressed?
   c. Identification of the problem's (waste's) root cause?

d. Identification and verification of the solution or improvement?
e. Standardization of the improvement, i.e., making it the new standard for the way the activity is performed?
f. Best practice deployment of the improvement, i.e., sharing it with related operations and activities that also might benefit from its implementation? The following list is suggestive but not mandatory.
    i. Where is the LMS's online database of completed improvement activities that allows keyword searches? How can a work group with a problem or improvement idea enter keywords into a search engine and find completed projects that might offer a solution or at least provide ideas?
    ii. When do work teams give project presentations to management and/or other work teams that might benefit from the information?
    iii. Where are there poster boards or other publicly accessible displays of completed improvement projects?

3. How does the CLPA ensure that projects, once they are initiated, proceed to completion and closure?
    a. Failure to carry all CLPA projects through to closure undermines people's confidence in the system and reduces their willingness to use it.
    b. "Closure" may include a decision that the proposed course of action is not feasible, desirable, or possible. This must, however, be a decision as opposed to neglect or inaction.

Provision 8.5.3: Preventive Action

LMS:2012 defers to the prevailing QMS standard on this item. Many forms of preventive action also constitute proactive action as discussed in the previous section.

DETAILS AND EXPANDED EXPLANATION

Chapter 9

Lean Management System: Details

A good quality management system (QMS) is a prerequisite for an effective Lean management system (LMS) and is therefore a mandatory provision of LMS:2012. The LMS must be able to take it as given that controlled procedures and work instructions exist; materials and test results are traceable; gages are calibrated; parts receive all necessary operations, tests, and inspections; and so on. In the absence of such a QMS, the system will probably be overrun with so much muda (waste) that any LMS will be ineffective. The choice of QMS is left to the user, although LMS:2012 assumes use of ISO 9001:2008 or the similar ISO/TS 16949 standard. Actual registration to ISO 9001:2008 or ISO/TS 16949 is optional, but the QMS does have to exist.

It is also recommended that the LMS be integrated into the QMS to facilitate the simultaneous audit of both. Ample precedents exist in the integration of ISO 14001 with ISO 9001, and now ISO 14001 with ISO 50001. As stated by Pinero (LRQA, 2011, 1–2),

> Because of the compatibility between [ISO 14001 and ISO 50001] they can very, very easily be integrated. The major benefit is that EnMS [energy management systems] and environmental management systems are so closely related in terms of the approach and the fact that they both have to deal with the environment and natural resources and so on.

Meanwhile, the Automotive Industry Action Group (AIAG) offers a course through its website on the integration of ISO 9001 and ISO 14001. This reinforces the principle that quality, environmental, energy, and even safety management are not four activities that take place on separate tracks but are instead synergistic and mutually supporting.

LMS elements that are not required by the QMS should, however, be identified clearly as such for the benefit of external QMS auditors. LMS:2012 is an unofficial standard for internal use in continuous improvement, and it is important to differentiate between its elements and those for which the organization is accountable to second-party (customer) or third-party (independent registrar) auditors.

To put this another way, ISO 9001 is sometimes described as "Say what you do and do what you say." What you say you do must meet the requirements of the prevailing standard, which is currently ISO 9001:2008. Then you have to do what you say (follow your own quality manual, second-tier procedures, and third-tier operating instructions) and be able to prove it from fourth-tier quality records; if it isn't written down, it didn't happen. If an LMS provision that is not required by ISO 9001 or ISO/TS 16949 is incorporated into the QMS, the organization has added an unnecessary element to its "say what you do" list for which it can be held accountable by external auditors.

One way to achieve this is to qualify each element in a general purpose business process manual with acronyms like QMS and LMS. For example,

> 4.1 (QMS): The organization *shall* implement, maintain, and continuously improve a quality management system consistent with the prevailing ISO 9001 standard…
> 4.1 (LMS): The organization *shall* create, document, maintain, measure, and continuously improve a Lean management system (LMS).

This tells external auditors that the first item is auditable under ISO 9001, but the second is not. Internal auditors should, however, audit both requirements simultaneously. This organizational approach can be broadened by qualifying certain LMS elements as LMS/E (energy, auditable under ISO 50001) and LMS/M (materials, in the context of environmental aspects, auditable under ISO 14001). In all cases, including quality, Lean, energy, and environmental aspects, a process perspective is critical.

Process Perspective

The importance of the process concept cannot be overemphasized. SIPOC is an important Six Sigma concept that encourages practitioners to look at business operations from the following perspective:

- *Suppliers* provide Inputs
- *Inputs* are resources for Processes. They may be physical items like parts and raw materials or services
- *Processes* are sequences of activities that transform inputs into Outputs
- *Outputs* are products or services that are delivered to Customers
- *Customers* use the processes' Outputs

Suppliers may be internal or external. A manufacturing process is the customer of the operations that precede it and the supplier of those that consume its outputs. *Processes do not necessarily stop at the company's boundaries*, and this brings up the close connection of SIPOC to the *supply chain*.

External suppliers that can't or won't deliver good quality, or whose inefficient operations add days or weeks to cycle times, can undermine the performance of the leanest company. An inefficient downstream distribution system can add enormous costs and wastes for the end customers. In simpler terms, you can't soar like an eagle if you flock with turkeys. Supplier and customer development are ways to turn the supply chain's turkeys into eagles.

Cianfrani and West (2010) underscore the importance of the process perspective. The process approach is important, and the systems approach recognizes the interaction of processes. *Most problems occur at the interfaces of processes*, e.g., during the handoff of work from one process to another. The QMS must address these interfaces very carefully for potential impacts on quality, whereas the LMS must examine them from the perspective of waste. Robinson (1990, 49) describes how work in process at a food packaging facility was carefully packaged and labeled, only to have the same packages torn open in the same room so the work could receive further processing. This wasteful exercise was apparently a requirement for the handoff of work from one administrative unit to another! Shingo (2009, 230) adds that an advantage of single-unit flow over processing by lot is the elimination of the need for boxes for transportation during the handoff process.

Consideration of value-adding motion versus waste motion also is extremely important. Henry Ford wrote long ago that pedestrianism is not a highly paying line of work, and observing or videotaping many jobs will often show just how much waste motion is built into them. It is difficult to pay an American worker $15 or $20 an hour to walk back and forth to get materials when offshore workers will do the same thing for $1 an hour. The best way to prevent jobs from going offshore is to make the American jobs so efficient that the per-unit labor cost ceases to be an issue even when wages are high.

Need for Documentation

There is a general misconception that QMSs like ISO 9001 are about paperwork and documentation. As with the QMS itself, the documentation is the servant and not the master.

Documentation is the foundation of all human progress. When people cannot read or write, their ability to teach and retain skills is limited to whatever they can transmit through oral tradition. What the trades once called knacks, or inventive tricks for doing jobs more quickly and efficiently, often died with their inventors.

It is therefore no coincidence that scientific progress and literacy both accelerated significantly after the invention of the printing press in 1440. Frederick Winslow Taylor speculated that trade workers had often found ways to improve the efficiency of their work throughout the centuries, but the improvements were lost because nobody wrote them down. There were probably two reasons: literacy was not widespread prior to the nineteenth century, and shop masters were unwilling to share secrets or knacks with competitors. Legends of "magic swords" probably arose from techniques such as hammering blades to introduce compressive stresses. *Metal Finishing News* (2009, website) says the secret of such swords, which could be bent almost double without breaking, was rediscovered only in the 1970s by means of x-ray diffraction. Toledo steel is certainly consistent with stories about Roland's sword Durendal, which the hero was unable to break to prevent its capture at the Battle of Roncesvalles.

The reference adds, "No other sword makers knew how to do this. The process was a closely guarded secret." Therefore, the masters and their journeymen would not have written a work instruction that a competitor or a foreign enemy might find, assuming that the masters and journeymen could write in an era in which literacy was limited primarily to the clergy. Inman (2006) adds that modern science cannot reproduce swords of Damascus steel, which were found to contain carbon nanotubes. "But since the techniques for making these swords have been lost for hundreds of years, no one is sure exactly why these swords are so exceptional." No work instructions exist for the manufacture of Damascus swords and, in the absence of reverse engineering, they might as well be magical even today.

Standardization and by implication documentation are easily explainable by the analogy to a ratchet, a mechanical device that allows a toothed wheel to turn in only one direction. Consider a labor-intensive task such as raising a bucket of water from a well, or using a capstan to raise a ship's anchor. If the crank or levers got away from the workers, the bucket would fall back into the well or the anchor would return to the seabed, thus wasting all the effort that had been put into the job. A ratchet made it impossible for the people to lose the work they had already put into the job. As shown above, Taylor speculated intelligently that improved ways to do jobs had been discovered and lost repeatedly. Ford and Crowther (1926, 82) then depicted standardization (and by implication documentation) as the ratchet that takes the "one step back" out of the "two steps forward and one back" model of progress:

> To standardize a method is to choose out of many methods the best one, and use it. Standardization means nothing unless it means standardizing upward.
> What is the best way to do a thing? It is the sum of all the good ways we have discovered up to the present. …Today's best, which superseded yesterday's, will be superseded by tomorrow's best.

A mechanical ratchet allows a load to move only upward, and standards along with documents ensure that a job's productivity can move only upward as well. The Ford Motor Company applied this concept to individual tasks very diligently, but this book has shown how quickly it forgot its Lean system and culture in the absence of a Lean manufacturing standard even in an era of widespread literacy, and even though Ford's books contain all the fundamental principles. This underscores the fact that, as with quality records, "If it isn't written down, it didn't happen."

Armies were the universal exception to the rarity of standardized and written instructions for all activities in pre-industrial times. They put enormous efforts into discovery of the best-known (i.e., fastest) ways to load muskets and cannons and then printed the results in drill manuals that everybody had to follow. Frank Gilbreth's observation of the motion efficiency of the drills led to his application of the underlying principles to civilian enterprises, with unprecedented gains in productivity.

Ford applied motion efficiency throughout his factories, and he stressed the importance of best practice deployment along with standardization. Standardization means that everybody performs a given job in the best known way, and this requires that the best known way be documented to prevent loss of the knowledge or backsliding to inferior methods. Best practice deployment means application of the knowledge in question throughout the organization, and this also requires documentation. Ford's statement (Ford and Crowther, 1926, 85) that "The benefit of our experience cannot be thrown away" summarizes the reason for ISO 9001's documentation requirements. Waste of experience and knowledge, as with waste of anything else, is not compatible with Lean manufacturing.

The next step is to describe the framework for the documentation. A typical QMS has four tiers:

1. A first-tier quality manual, which often follows the standard to which the QMS will be audited (e.g., ISO 9001, ISO/TS 16949).
2. Second-tier documents define *policies* or *operating procedures* for organization-wide activities like record retention and document control. These must support the relevant sections of the quality manual.
3. Third-tier *work instructions* or *operating instructions* are for specific jobs and operations.
4. Fourth-tier *quality records* include maintenance and quality inspection logs and also closed loop corrective actions.

A four-tier structure is not mandatory. Dobb (2004, 26) says that a small organization can often make do with a three-tier system in which any necessary work instructions are absorbed into the policies or operating procedures. Work instructions are often unnecessary for skilled trades and professions. In contrast, a very large operation might require five or more tiers, but it is recommended explicitly that the hierarchy be kept as simple as possible.

Every document in the system is subordinate to the one above it. A maintenance log is, for example, an implied instruction to do preventive maintenance, but a quality record does not have the independent authority to tell anybody to do anything. A work instruction must define the preventive maintenance activity, and it also must say to record completed work in the maintenance log. The information that the maintenance log calls for must be identical to that which the work instruction says to record because conflicts between superior and subordinate documents are nonconformances under ISO 9001 and similar QMS standards. The need to coordinate the content of all documents in the system cannot be overemphasized, as poor document control is a principal cause of nonconformance to the QMS standard (Bakker, 1996).

In many if not most cases, the Lean system documentation will overlap with the quality system documentation. As a simple example, maintenance logs (which support total productive maintenance) are quality records that are already subject to control under ISO 9001 and ISO/TS 16949-compliant systems.

Lean Manual

The concept of synergy between Lean and quality cannot be overemphasized. Piggybacking an LMS onto a QMS should not require a lot of extra work because many of the techniques and principles are synergistic and mutually supporting.

Any existing QMS will almost certainly include Lean methodology. Design for Manufacture, error proofing (poka-yoke), autonomation (jidoka), visual controls, work standardization, best practice deployment, team-oriented problem solving, preventive maintenance, consideration of the work environment (temperature, humidity, lighting, vibration, and so on), and many other techniques support both quality assurance and waste reduction.

Many forms of waste (as defined in Lean manufacturing) also create opportunities for poor quality, which is itself a form of waste. The waste of inventory seems harmless enough because it is hard for a nonperishable product to acquire defects by sitting in a warehouse or stockroom. The truth is, however, that inventory gives defects a place to hide instead of forcing them to become immediately visible. Inventory reduction will therefore improve quality even if the immediate goal is to reduce carrying costs and cycle time.

Remember, however, that the quality system is primarily concerned with conformance to customer requirements, such as specifications. It is quite possible to deliver perfect quality while wasting enormous quantities of time, energy, and materials, and these issues should be the primary focus of the Lean manual.

Control and Retention of Documents and Records

The need to keep obsolete and conflicting documents out of the system cannot be overemphasized. Recall that Bakker (1996) cites document control as a major source of audit findings, and this makes sense because so much can go wrong with it. Smith (2011a) adds, "It is commonly reported that document control generates more nonconformances than any other quality management system (QMS) element," although he adds that this does not have to be the case unless organizations make it more complicated than it needs to be.

Preservation of records also cannot be overemphasized, noting again that "if it isn't written down, it didn't happen," with "it" referring to something like preventive maintenance, a required test or inspection, gage calibration, or something similar. The author had the experience of a brief power failure while saving a financial spreadsheet record, with the result that the entire spreadsheet file was wiped out. Fortunately, there was a backup on a flash drive, so only about 15 minutes worth of work was lost.

This is a strong argument for saving quality records on two disk drives, which are preferably not in the same building so they cannot be destroyed simultaneously by a fire, flood, or other disaster. Commercial backup services are available, and records can also be saved on CD-ROMs and similar media for offline storage.

Chapter 10

Organizational Responsibility: Details

Need for Organizational Commitment

A Lean manufacturing or Lean enterprise initiative stands or falls on management and workforce commitment. Implementation of world-class Lean manufacturing requires upper management to agree to a no-layoff policy, whereas labor must in turn repudiate all restrictive work rules. These requirements will doubtlessly encounter resistance from management and labor unless both understand the reasons and the mutual benefits. This section and the ones that follow provide the implementers with information with which to gain buy-in from both entities. Ford and Crowther (1922, 117) *summarize the entire science of industrial and labor relations in a single sentence*: "It ought to be the employer's ambition, as leader, to pay better wages than any similar line of business, and it ought to be the workman's ambition to make this possible."

Labor, and especially unionized labor, must realize that no system can pay more in wages and benefits than it creates in value. If labor demands more value than it produces, the jobs will soon cease to exist or they will be sent offshore. If management seeks to pay less than the jobs are worth, it should not be surprised when labor delivers as little work and effort as it can. Frederick Winslow Taylor (1911a, 8) made this emphatically clear more than 100 years ago:

> ...after a workman has had the price per piece of the work he is doing lowered two or three times as a result of his having worked harder and increased his output, he is likely entirely to lose sight of his employer's side of the case and become imbued with a grim determination to have no more cuts if soldiering can prevent it.

Soldiering (marking time, like soldiers marching in place but going nowhere) is the deliberate and systematic limitation of output to prevent cuts in piece rates or by implication layoffs. The workforce applies peer pressure to rate busters, or employees who produce more than the other workers have agreed informally to produce, because labor does not want management to know that greater productivity is possible. The supplemental information for document control pointed out that knacks, or better ways to do jobs, are frequently lost because nobody writes the new method down. Soldiering means workers hide any knacks they have discovered or else use them to make their agreed work quota easier without increasing the overall output.

Layoffs are, of course, the equivalent of piece rate cuts and are equally likely to make the worker "lose sight of his employer's side of the case." This book has already cited Upton Sinclair's (1937, 81) observation that Ford's successors or stewards found ways to circumvent a rule against firing workers when productivity improved; they did not discharge the workers outright but instead harassed and rode them until they left or until (possibly) a performance-related issue could be fabricated against them.

Layoffs, whether overt or by pressuring employees to leave as described by Sinclair, are totally inconsistent with Lean organizational culture. The same applies to union contracts that demand a certain number of jobs (and presumably dues-paying union members). If, for example, productivity improvements allow 800 employees to do the work of 1,000, the organization must not lay off 200 people. If it does, the remaining workers will rightfully withhold their cooperation from further improvement efforts. If 25 people retire that year, however, the union must not demand that their "places" be filled with new employees.

Meanwhile, Levinson (2009) points out that General Motors' (GM) infamous "job bank," in which unnecessary employees received full pay for showing up to work crossword puzzles, watch videos, or at best take classes and perform community service, was the wrong thing for the right reason. "Right reason" means that GM did not lay people off when they became unnecessary, so the union had no reason to soldier or suppress productivity-improving knacks. "Wrong thing" means the workers should have been assigned to value-adding work instead of undignified idleness or make-work. China had recently delivered millions of defective valve stems to U.S. automakers, and, because the marginal cost of assigning their production to these idle American workers would have been zero, that is what should have been done.

"Marginal cost of zero" means GM had to pay the workers whether it used them or not. The incremental cost of the valve stems would have therefore consisted only of the materials plus any required energy. The material and energy costs would have been roughly identical for China, and the cost to import the valve stems would have been eliminated along with the ability of defects to hide in inventory.

The idle workers can also look for ways to use the company's existing technology to fill new market needs. Toyota now faces an aging Japanese population and less demand for its automobiles. It is therefore adapting its automotive production technology to meet the senior citizens' needs, such as hydraulic home elevators and cars that can accommodate wheelchairs (Sapsford, 2005).

Another way to avoid layoffs is to use overtime instead of hiring more workers who may later become unnecessary. Overtime is nominally 50 percent more expensive but actually less so because certain benefits like health insurance do not change. This does not mean overtime should be used routinely; if it is, then market demand is probably consistent enough to justify more full-time workers. Yet another procedure is to use temporary workers, who do not expect permanent employment, to take routine or low-skill tasks from the regular workforce during times of increased demand. Always remember, though, that any of these approaches—overtime, more full-time workers, and temporary workers—is inferior to making jobs more productive so they can handle the increased workload with the same number of workers.

Management Commitment Loses the Luddites

The fear of machinery putting people out of work predates the Luddites, English textile workers who smashed weaving machinery, by centuries. James and Thorpe (1994, 388–389) report that Rome's Gallic provinces developed a reaping machine known as a vallus. It was something like a wheelbarrow that a donkey pushed from behind while a comb or set of teeth in front guided the crop into knives that cut the stalks to drop the harvest directly into the wheelbarrow. The reference says that these machines never caught on because Rome feared social upheaval if they put slaves out of work. Nobody seemed to realize that it was doubtlessly cheaper to pay a couple of men to run a vallus than it was to feed and house the equivalent amount of unpaid labor necessary to do the same work with scythes.

The reference also reinforces the importance of documentation, as in "If it isn't written down, it didn't happen" (James and Thorpe, 1994, 388–389). The concept of harvesting machines was forgotten until the nineteenth century, when the inventor John Ridley read about the vallus thanks to the writing of a fifth-century Roman named Palladius. "A translation and reconstruction drawing was published in J.C. Loudon's *Encyclopedia of Agriculture* in 1825," and this is what inspired Ridley to reinvent this labor-saving device (James and Thorpe, 1994, 388–389).

The Luddite mentality has unfortunately survived into the twenty-first century. Levinson (2002b, website) describes how the International Longshore and Warehouse Union (ILWU) objected to the Pacific Maritime Association's (PMA's) desire to automate logistics information processes with bar code scanners. "But ILWU wants to continue entering shipment data by hand instead of scanners. The ILWU hopes to protect jobs from the effects of automation." Recall also that Shirouzu (2001) reported that a Ford worker resisted the removal of 2,000 footsteps from his daily work because he was afraid the increased efficiency would result in layoffs.

Simple mathematics shows that Ludditism destroys jobs and does not protect them. Recall that Frank Gilbreth's non-stooping scaffold allowed each worker to lay 350 bricks per hour with less effort than he previously required to lay 125. This example can now be turned into a simple 15-minute classroom exercise for management and labor alike, to reinforce the need for both management and workforce commitment.

Suppose for example that the customer would pay two cents for every 10 bricks laid (over and above the cost of the bricks, mortar, animal or rail transportation, and so on) in the money of the early twentieth century. From this, we know that 1,750 bricks per hour would come to $3.50 per hour that can be paid as wages. This means that 14 men who each lay 125 bricks per hour can get 25 cents per hour, or $2 for eight hours of exhausting labor in which they bend over 125 times per hour to pick up each brick. Five men who use Gilbreth's non-stooping scaffold to lay 350 bricks per hour can get 70 cents per hour, or $5.60 for 8 hours, under the same circumstances. This would have easily put the bricklayer into the middle class during that era.

However, labor must realize that 14 men cannot be paid 70 cents per hour if they lay only 125 bricks per hour due to either soldiering or lack of the technology with which to lay 350 per hour. Management must understand, in turn, that the workers are not going to lay 350 an hour, even with the non-stooping scaffold, if they continue to receive only 25 cents per hour. If the organization wants to employ 14 men at high wages, it must offer a discount that will let it sell the laying of 4,900 bricks per hour. This will mean slightly lower wages (albeit still far higher than under the old method) but full employment for all 14 workers. The construction firm will indeed sell its full capacity at the discounted price if its competitors do not adopt the new labor-saving methods and compensate their own workers accordingly. Emerson (1909, 19) elaborates on this example substantially, and comes to the same conclusion:

> That men should work very hard for 9 or 10 hours per day is not a hardship if they are interested in their work, or if, in the larger interest of the community, they work efficiently; but to work desperately hard for many hours at dirty, hot and rough work, yet waste 67 per cent of the time and effort, is unpardonable. What could have resulted from an elimination of this waste?
>
> 1. The product could have been cheapened.
> 2. The men could have worked one-third the time and have accomplished as much.
> 3. One man could have done all the work and have earned three times as much.
>
> The benefits should however be distributed in all three directions. Fewer men should work less hard, receive higher wages, and deliver a cheaper product.

The last sentence describes exactly what Henry Ford sought to achieve and did achieve during the next decades: higher wages, lower prices, and higher profits simultaneously. Cost reductions from Lean manufacturing therefore increase sales and create additional work for workers whom improvements leave temporarily idle. Ford's experience was that the lower automobile prices that his methods made possible increased sales to the extent that he had to hire more workers despite enormous efficiency improvements.

Management and Workforce Commitment: Workforce Flexibility

Restrictive work rules that say for example that only a press operator may touch a press and only a lathe operator can touch a lathe are totally inconsistent with a Lean management system (LMS). Ford and Crowther (1922, 94) describes the culture that must prevail in a Lean organization as follows: "The health of any organization depends on every member—whatever his place—feeling that everything that happens to come to his notice relating to the welfare of the business is his own job."

Workers who are temporarily idle for lack of work in their own job must be willing and able (i.e., cross-trained) to work at something else, as was the practice at the Detroit, Toledo, and Ironton (DTI) railroad.

> The divisions of work among the men were abolished; an engineer can now be found cleaning an engine or a car or even working in the repair shop. ...The idea is that a group of men have been assigned to run a railroad, and among them they can, if they are willing, do all the work. (Ford and Crowther, 1926, 200)

Japanese workers are often cross-trained on many jobs, and they will "follow the work" around their factories. If their own workstation has nothing to do, they will go to one that has excess work. If there is nothing else to do, they will look for ways to make their jobs more efficient and more productive.

Productivity improvements may indeed eliminate jobs, and the workers in question must be willing and able to do something else. The new work will usually be more interesting and satisfying than the old because Lean manufacturing eliminates drudgery and waste. No employee can afford to regard himself or herself as being fixed in one line of work, and everyone must be willing to learn new tasks.

Management Commitment and Training

The importance of management commitment to training cannot be overemphasized. George (2002, 19–20) describes how a CEO hired W. Edwards Deming to teach Total Quality Management to senior executives. The CEO told the executives how important the training was, and that he expected their diligent participation, but then he left the room. Some of the managers therefore substituted a quick reading of the cover of a videocassette of Deming's lectures for actual viewing of its 16 hours of content. In other words, even the reputation of W. Edwards Deming was not enough to gain buy-in by management professionals after the CEO's own demonstrated lack of interest. Compare this example to that set by Field Marshal Aleksandr V. Souvaroff as depicted by Lord Byron (1857) in *Don Juan* (canto 7):

> Glory began to dawn with due sublimity,
> While Souvaroff, determined to obtain it,
> Was teaching his recruits to use the bayonet.
> It is an actual fact, that he, commander
> In chief, in proper person deign'd to drill
> The awkward squad, and could afford to squander
> His time, a corporal's duty to fulfil...

Byron therefore paints Souvaroff as an eccentric commander who performs the work of a drill sergeant, which was in fact one of his success secrets. Souvaroff said effectively that training was the most important thing the Russian Army did, and his personal participation ensured that his subordinates took this statement seriously. The superior training of Russian soldiers made Souvaroff's army an unstoppable war machine that crushed anybody against whom the Tsarina or Tsar sent it, including many of Napoleon's future marshals and the very skilled Polish commander Tadeusz Kosciuszko. Kosciuszko had not only played a major role in the United States' eventual victory in the War of Independence, he defeated every other Russian commander he fought.

In summary, LMS:2012's Management Commitment and Workforce Commitment sections are designed to promote an environment that complies with the following principle:

> Scientific management... has for its very foundation the firm conviction that the true interests of the two [employers and employees] are one and the same; that prosperity for the employer cannot exist through a long term of years unless it is accompanied by prosperity for the employee, and vice versa; and that it is possible to give the workman what he wants—high wages—and the employer what he wants—a low labor cost—for his manufactures. (Taylor, 1911a, 1)

Lean Management Policy

Here is a one-sentence sample Lean policy that also supports ISO 14001 and ISO 50001: "It is the policy of XYZ Corporation to remove all identifiable wastes of (1) time of the product or service, (2) time of people, (3) materials, and (4) energy from its supply chain."

This is easy for employees and other stakeholders, including customers and suppliers, to remember. It promotes the desired results by encouraging people to focus on a small number of very specific and easily understandable key performance indicators (KPIs) instead of generic instructions to "eliminate waste." An employee who finds himself or herself walking to get parts or, even worse, waiting at a stockroom will immediately identify this as "waste of the time of people." An employee who sees discarded packaging materials will say immediately, "This is a waste of materials, unless those containers can be sent back for another load or else carry one of ours." As with the quality policy, the Lean policy must not be a nice-sounding slogan but rather a concise set of principles that drive observation, action, and results.

Supply Chain Responsibility

Recall that Ford and Crowther (1922, 94) said quite rightly that "The health of any organization depends on every member—whatever his place—feeling that everything that happens to come to his notice relating to the welfare of the business is his own job." The well-known story about Somebody, Everybody, Anybody, and Nobody shows, however, that a job that is Everybody's responsibility soon becomes Nobody's responsibility.

It is fairly easy to reconcile these two apparently conflicting observations through a combination of empowerment and clearly defined responsibilities. The organization trains and empowers Everybody to identify all forms of waste and to initiate closed loop corrective action (or proactive improvement) to remove it. Anybody is free to initiate an error cause removal (ECR) or scare report (hiyari) for a quality or safety problem or for any form of waste.

However, a particular Somebody must be responsible for assessment of each process or operation for waste, possibly through the recommended KPIs. The most rational choice is the process owner, who would be the project champion in a DMAIC, TOPS-8D, or CQI-10 corrective action or proactive improvement activity. The management representative is the specific Somebody who is responsible for the overall LMS. This ensures that Everybody's job, which Anybody can do, will not end up becoming Nobody's job.

Responsibility for the outcome of a task requires worker authority (or autonomy) in the form of what Juran and Gryna (1988, 10.27–10.31) define as a state of worker self-control.

State of Self-Control

1. Workers know what they are *supposed to do.* "Goals and targets are visible" (p. 10.28); the workers are aware of the process's intended outcomes and what constitutes good or bad quality. Knowledge of how to do the work, in the form of work instructions, training, drawings, visual aids, and so on, is also available.
   a. The latter consideration reinforces the need for controlled documents.
2. Workers know what they *are doing.* Feedback is available to the workers and preferably with no lag time, as in the case when the work must go offline for measurement in a laboratory or quality department.
   a. This is compatible with *self-check systems* that reject nonconforming work before it can leave the workstation.
3. Workers *can regulate* the work. This means they are empowered to correct abnormalities and defects instead of having to ask a supervisor or technician what to do.
   a. As an example, the out-of-control action procedure (OCAP) or corrective and preventive action matrix (CP matrix) tells operators what to do if a statistical process control chart shows an out-of-control signal.
   b. "Regulation" should, however, go beyond the mere correction of abnormal conditions and defects. The LMS must empower frontline workers to implement (with appropriate project controls) permanent solutions to prevent the problems from recurring.

Optimization of an LMS or a quality management system (QMS) for that matter meanwhile relies on effective two-way communications throughout the organization and indeed the entire supply chain. The importance of communications cannot be overemphasized.

Supply Chain Communications

All human progress relies on communication, and this book's emphasis on documentation and document control is but one example. The fact that ancient civilizations developed around rivers, and exclusively around rivers, is particularly instructive. The rivers in question were the Nile (Egypt), Tigris and Euphrates (Mesopotamia), Ganges and Indus (India), and Yellow River (China). Rivers provide water for agriculture, which allows more people to live in a single place and exchange information. They are also an avenue of navigation, which facilitates the exchange of information between people who live in different places.

It is also significant that seafaring nations often tended to advance more rapidly than landlocked ones, again because of communication with people in distant lands. Intelligent seafarers benchmarked their own country's practices against those of people with whom they came into contact, and adopted superior or innovative practices. Maize, the potato, cotton, and unfortunately tobacco were imports that Europeans discovered in other lands and purchased for their own use.

The information in question could be transmitted orally, e.g., through sailors' stories that often changed and grew with the telling, or more ideally by writing. The discussion on document control emphasized the fact that scientific knowledge began to propagate very rapidly after the invention of the printing press. It was nonetheless necessary for a scholar to find a book with the particular knowledge he or she wanted, with membership in an academy of science being perhaps the only way to find books or information quickly. Library and information science, which groups books by subject matter, later made the desired books easier to find.

Internet search engines with keyword search capability are the latest breakthrough that allows easy discovery of information. This suggests that human knowledge will grow even more rapidly in the twenty-first century than it grew in the twentieth. This observation reinforces the desirability of making documents keyword searchable by everybody in the organization. This includes quality records like corrective action requests and proactive improvement project reports.

The next step is to apply this need to exchange nonconfidential information easily with every element of the supply chain, and this requires what management expert Tom Peters calls porosity. Porosity is the absence of internal and external barriers to communication and collaboration. Meanwhile, one of W. Edwards Deming's 14 Points is to break down barriers between internal departments, i.e., eliminate the functional silo effect.

Need for Internal and External Porosity

The individual who has his or her hands on a specific task, and this may include a production worker in a customer's factory, knows more about the needs of that task than anybody else. It is therefore important for that person to be able to communicate with the supplier, preferably without the need to go through "channels" (other than on an awareness basis) to do so. Ford and Crowther (1922, 91) elaborate,

> If a straw boss wants to say something to the general superintendent, his message has to go through the sub-foreman, the foreman, the department head, and all the assistant superintendents, before, in the course of time, it reaches the general superintendent. Probably by that time what he wanted to talk about is already history.

This principle applies as much if not even more so to communications between supply chain partners. Levinson and Tumbelty (1997, 16) describe customer contact teams (CCTs) that include frontline production workers who visit the customer's gemba (value-adding workplace) to see how the customer uses the product and get ideas from the customer's workers as to how to improve the product. The CCT is one medium for making an organization porous to other parts of the supply chain.

Lean System Review

Management review of the system plays an important role in best practice deployment by looking at localized findings and observations from an organizational perspective. Japan's *policy diagnosis* is similar to best practice deployment, but it applies specifically to review of the effectiveness of a QMS or, by implication, LMS, environmental management system, or energy management system.

The basic idea is that one termite (quality system nonconformance or waste) is generally symptomatic of a termite nest, and that stepping on the single termite does not eliminate the underlying root cause of the problem. Suppose, for example, that an auditor finds an obsolete document or out-of-calibration gage in a work area, which is a quality system nonconformance. Replacing the document or calibrating the gage does not in any way answer the question as to how the nonconformance got there in the first place. We must ask instead, "How did our document control system allow an obsolete document to remain in a work area? How did our calibration management system allow a gage to miss its calibration?" The next step is to see if the solution or improvement might apply to other activities or processes, which is best practice deployment.

Chapter 11

Infrastructure and Resources: Details

A manufacturing or service operation requires appropriate human, physical, and information resources to function correctly. This chapter provides details and ideas.

Workforce Training and Empowerment

The phrase "it worried the men" with regard to waste at Ford's River Rouge plant (Norwood, 1931) sums up a key element of a successful Lean organizational culture; the workforce must be able to identify all forms of waste and be unwilling to accept any of them.

Many forms of waste differ from poor quality by being asymptomatic. A quality problem triggers the organizational equivalent of pain: scrap, rework, or a customer complaint. This draws attention to the problem and triggers a corrective action request or quality action request. A process can deliver perfect quality even when other forms of waste like excessive inventory, non-value-adding transportation, and waiting time are present.

Bending over to pick up each brick from the ground, with the mason lowering and raising his entire upper-body weight each time, was the generally accepted method of bricklaying until Frank Gilbreth introduced the non-stooping scaffold. The waste was *asymptomatic* because the walls were getting built, but eliminating the waste allowed them to be built more quickly and with far less physical effort.

Another common characteristic of many forms of waste is that people become used to working around them or working despite them. A very simple way to visualize this concept is to imagine a machine that stops frequently but can be restarted with a swift kick. Kicking the machine a few dozen times a day becomes an accepted part of the job, and the waste of its so-called minor stoppages soon becomes built into the job.

The ability to recognize waste on sight and to teach this skill to others was among Ford's principal success secrets. The following statements underscore the fact that *waste must be identified before it can be eliminated*:

- "It is the little things that are hard to see—the awkward little methods of doing things that have grown up and which no one notices. And since manufacturing is solely a matter of detail, these little things develop, when added together, into very big things" (Ford and Crowther, 1930, 187).
- "Unfortunately, real waste lurks in forms that do not look like waste. Only through careful observation and goal orientation can waste be identified. We must always keep in mind that the greatest waste is the waste we don't see" (Shigeo Shingo, quoted in Robinson, 1990, 14).
- "In reality, however, such waste [waiting, needless motions] is usually hidden, making it difficult to eliminate. …To implement the Toyota production system in your own business, there must be a total understanding of waste. Unless all sources of waste are detected and crushed, success will always be just a dream" (Ohno, 1988, 59).
- *Friction* is "…the force that makes the apparently easy so difficult. … countless minor incidents—the kind you can never really foresee—combine to lower the general level of performance, so that one always falls short of the intended goal" (Clausewitz, 1976, 119).
- "The accumulation of little items, each too trivial to trouble the boss with, is a prime cause of miss-the-market delays" (Peters, 1987, 323).
- "They turned out to be the little things that get under a worker's skin but are never quite important enough to make him come to management for a change" (Halpin, 1966, 60–61).

Lean management system (LMS) auditors should therefore ask *how* the organizational culture promotes zero tolerance of little items "too trivial to trouble the boss with," "little things that are hard to see," or "countless minor incidents" that undermine the organization's performance.

Facilities, Layout, and Supporting Services

Ford's introduction of separate electric motors for production equipment made the modern U-shaped work cell possible. Power had previously come from overhead shafts with pulleys that could be connected to each tool, and this dictated straight process layouts. It is important to note that a process-oriented, departmental, or "farm" layout that places all machines of a similar type in a single area of the factory requires interdepartmental transportation. This, in turn, requires batching of work for transfer, which increases inventory and works against single-unit processing.

In many cases, however, utility-intensive equipment makes it difficult to rearrange the tools into ideal work cells. As an example, an etching station that uses and must dispose of hazardous fluoride wastes cannot be moved around at will like a medium-sized drill press. The same practical considerations apply to chemical plants and other continuous-process industries, although their flow nature is automatically conducive to Lean operation.

Work Environment, Ergonomics, and Motion Efficiency

Henry Ford's job design guidelines said that no task should ever require a worker to take more than one step in any direction to fetch or transfer materials; pedestrianism is not a highly paying line of work. No job should require a worker to bend over; it is not good for the employee's back and it adds his or her entire upper-body weight to whatever is being lifted.

No job should require a worker to routinely apply significant physical effort (e.g., turning wrenches or screwdrivers). It is inefficient, and it increases the risk of repetitive-motion injuries. Many jobs at Ford's plants were extremely repetitive, with workers performing the same motions hundreds or even thousands of times a day. There were, however, few if any reports of repetitive motion injuries because of this job design principle.

Cleanliness at the Ford plants was promoted by having waste containers within seven steps of any location on the shop floor. A similar practice was later adopted by Disney theme parks, and that is why they are usually free of litter.

Gilbreth (1911, Chapter 3) points out, "Light in a factory is the cheapest thing there is. It is wholly a question of fatigue of the worker. The best lighting conditions will reduce the percentage of time required for rest for overcoming fatigue. The difference between the cost of the best lighting and the poorest is nothing compared with the saving in money due to decreased time for rest period due to less fatigued eyes."

Fluorescent light is not the white light (as looking at it through a prism will show) with which the human eye is designed to work. Its production of greenish-tinged images, as often seen in manufacturing trade journals, is caused by its interaction with daylight film that is designed to work with white light. (Good photographers know how to eliminate this effect with filters, and digital cameras can take care of the problem automatically today.) These facts plus Gilbreth's observations raise the very real possibility that the fluorescent lights that are used in most factories are a false economy. As with most other forms of waste, however, people have learned to live with the waste (eye fatigue) that comes from the use of fluorescent lights that were selected for energy efficiency.

Chapter 12

Product or Service Realization: Details

Planning

When most people think of Lean, they think of kaizen blitzes, workplace rearrangements, implementation of error-proofing devices, and so on. All these techniques suppress waste in the process at hand, but they can rarely eliminate waste that was designed into the product and/or process in the first place. The principle is similar to the one that says it is easier to design quality into the product than it is to manufacture it into a bad design. It is necessary to be proactive from the very start of the product or service realization process, and therefore in the design and development phase.

Design and Development for Lean

Design for Manufacture (DFM) was a key aspect of Henry Ford's manufacturing strategy:

> Start with an article that suits and then study to find some way of eliminating the entirely useless parts. This applies to everything—a shoe, a dress, a house, a piece of machinery, a railroad, a steamship, an airplane. As we cut out useless parts and simplify necessary ones, we also cut down the cost of making (Ford and

> Crowther, 1922, 14). But also it is to be remembered that all the parts are designed so that they can be most easily made (Ford and Crowther, 1922, 90).

DFM included not only designing parts for ease of manufacture but also to minimize the need for machining which generates waste, as shown by the key performance indicators (KPIs) for waste of material and energy. Ford minimized this waste by stamping or forging several pieces to near-final shape and then welding them together rather than casting a large piece and then machining large quantities of metal from it. *Any process that produces huge piles of metal shavings (and used cutting fluid) should be questioned immediately as a source of material and energy waste.*

It cannot be emphasized too much that waste is easy to overlook, especially when the product's quality is good. Most people would not give a second thought to huge piles of shavings from machining operations as long as a satisfactory product is going to the customer. A paradigm shift is necessary to get people to notice this kind of waste, and Levinson (2005) recommends a *Bill of Outputs* (BOO) that compels accountability for not only the product but also the materials that are expended in its creation.

Purchasing and Lean Supply Chain Practices

It *may* be acceptable to buy huge quantities in advance of needs to get a substantial discount as long as the material is not subject to deterioration or obsolescence. Ford kept huge inventories of coal and iron ore for his steel mills, which was contrary to his practice of buying only for immediate needs. As a practical matter, however, both commodities had to be transported in bulk. Furthermore, obsolescence was not an issue because their eventual use was a certainty.

The same concept applies to customizable inventory. Companies like Zazzle, CafePress, and so on keep inventories of generic T-shirts, coffee mugs, and so on that can be printed to order. The integrated circuit industry can meanwhile produce master slices, or silicon wafers with embedded semiconductor devices. The master slice is a relatively safe form of inventory that is unlikely to become obsolete because it can be personalized into a variety of end products by wiring the devices in different ways.

Purchasing Process

The adage that you can't soar like an eagle when you work with turkeys is very applicable to supply chain management. Subcontractors with long lead times for turning work around can undermine the effectiveness of the Leanest company, as described by Womack and Jones (1996, 71): "Bumpers disappeared into Chrome Craft [a subcontractor for a bumper manufacturer] and didn't reappear for weeks." The bumper manufacturer's owner and a Toyota *sensei* (teacher) practiced supplier development by teaching Chrome Craft how to do Just-in-Time (JIT).

It is also is hard to reduce costs when suppliers and subcontractors keep raising prices. Ford and his production chief Charles Sorensen used supplier development when dealing with the C.R. Wilson Body Company, which wanted $152 per automobile body. Sorensen showed that the body could be built for $50 in labor and materials, and Ford then offered C.R. Wilson $72 per body. The $22 in profit was more than the supplier would have made by using its own (non-Lean) methods and "…the upshot of it was that he made more money out of the low price than he had ever made out of the high price, and his workmen have received a higher wage" (Ford and Crowther, 1926, 43–44).

Lean Production and Service

The assessment questions for Lean production and service (Chapter 7) included Lean production control systems such as kanban and Drum-Buffer-Rope (DBR), and these techniques extend to activities other than manufacturing. Ford used heijunka (production smoothing) to largely eliminate rush hour traffic from the River Rouge plant even though it employed more than 100,000 workers (Norwood, 1931, Chapter 3). Arrival and departure times were staggered so that no more than 8,000 arrived at any given time, which was in fact a major improvement over the previous figure of 14,000. The superiority of small lot production over large runs was therefore applied successfully to the problem of traditional shift changes that would have otherwise involved the simultaneous arrival and departure of more than 30,000 workers.

The same principle could, for example, be applied to airport shuttle buses. If a car rental company's bus circulates every 20 minutes, it will drop off a large load of passengers in a previously empty rental office with the result that customers will have to wait while rental agents work at 100 percent capacity. Suppose, however, that two rental agencies with neighboring locations pooled their buses to pick up and drop off passengers every 10 minutes. This would smooth the workload at both offices, reduce customer waiting time, and also reduce the "feast or famine" effect on the workers. This idea came to mind when the author saw a shuttle bus with two car rental companies' names on them, although the reason was that the companies had merged as opposed to pooling their buses. The lesson is that heijunka is quite applicable to service as well as manufacturing activities.

Lean Process Control

The basic concept of a visual control is that it makes the status of an activity easily visible to anyone who is observing it. Andon lights show at a distance whether a piece of equipment is working, idle, or down for maintenance. Empty kanban squares are an immediate signal to the upstream operation to produce more material, whereas full ones tell the upstream operation to stop. Caravaggio (1998, 134–135) summarizes visual controls as follows:

> Visual controls identify waste, abnormalities, or departures from standards. They are easy to use even by people who don't know much about the production area.
>
> A visual control system has five aspects:
>
> 1. Communication: Written communications are easily accessible.
> 2. Visibility: Communication with pictures and signs.
> 3. Consistency: Every activity uses the same conventions.
> 4. Detection: There are alarms and warnings when abnormalities occur.
> 5. Fail-safing: These activities prevent abnormalities and mistakes.

Procedures, and preferably built-in and automatic ones, to keep nonconforming work out of the process also constitute Lean process controls. Shingo (1985, Chapter 5) cites:

1. *Successive check systems*, in which a worker other than the one who made the part inspects it. This can be the worker who receives the part for the next process, in which case the successive check is effectively

an incoming inspection by the internal customer. Shingo stresses the fact that prompt feedback to the process that makes a defect is vital. This does not unfortunately change the fact that any kind of inspection whose purpose is to protect a customer, whether internal or external, from nonconforming work is value-assisting at best.

2. *Self-check systems* do not wait for the work to reach the next process (internal customer), and they can often be automated. Automatic sorting gages can automatically reject parts that do not meet specification, and they also use a visual (or audible) signal to tell the worker that the machine has produced a nonconforming part.
3. *Autonomation* (jidoka) means the equipment can detect an abnormal condition and either correct it or alert the worker. The reference's chapter does not mention it explicitly, but an example of a self-check system (Shingo, 1985, 81) shows a machine that can detect broken drill bits, shut itself down, and activate a flashing red light (andon light, another visual control) to alert the worker. It achieves this with a verification device that moves in tandem with the drill, and attempts to insert verification rods through the most recently produced parts. If the rod cannot go through, the hole has not been drilled.
4. *Source inspections* eliminate defects by preventing abnormal conditions, or at least preventing them from making nonconforming work.
   a. *Vertical source inspection* means control of the incoming work and upstream processes to exclude defects or nonconformities. The focus is on process controls as opposed to incoming inspections.
   b. *Horizontal source inspections* seek to detect abnormal conditions before they can produce nonconforming work. Shingo (1985, 86–87) cites a vacuum cleaner packaging operation that made it impossible to leave out parts or instruction booklets. The worker could not send the box on unless he had picked up every necessary part (as detected by a limit switch), which is a form of poka-yoke (error proofing).

Horizontal source inspections also apply to many health care activities. HyGreen's Hand Hygiene Recording and Reminding System, which can also be described as error proofing, makes it impossible for health care workers to forget to wash their hands, just as the previously described setup made it impossible for workers to forget to leave parts out of the vacuum cleaner package. When the employee washes his or her hands with soap or antiseptic gel, a detector senses the soap or gel and transmits this information to a special badge worn by the employee. A light on the badge turns green to indicate that the employee is now ready to work with a patient. If the employee approaches the patient's bed without washing his or her hands, the badge vibrates to remind the worker. PRWeb (2011) reports that the system was sufficiently effective to reduce hospital-acquired infections by 89 percent at Miami Children's Hospital.

Poka-yoke means, of course, error proofing in general. Shingo had originally used baka-yoke (idiot proofing), but workers took offense to this. "Error proofing" is a far better term because, if it is possible to do the job the wrong way, it is only a matter of time until the even most intelligent and diligent worker does it the wrong way.

The Ford safety principle "Can't rather than don't" (Norwood, 1931, 83) is related very closely to poka-yoke, and it supports industrial safety and hygiene. It means that, instead of relying on worker vigilance to avoid accidents, the job makes it impossible for the worker to have an accident. Resnick (1920, 8) shows how the Ford Highland Park plant applied "Can't rather than don't" to a punch press: "To trip a press equipped with this device, the operator must press two buttons, about a foot apart, so that it is necessary for him to use both hands." It was therefore impossible for the worker to have a hand in the press when it closed. A two-worker press required the simultaneous depression of four buttons for the same reason.

Supply Chain Management

The concept of supply chain management cannot be overemphasized because true Lean operations require coordination and cooperation between supply chain partners. Ford's vertically integrated system—that is, Ford owned not only the automobile factories but also most of the raw material sources along with the logistics systems that delivered them—turned the business into what Norwood (1931, 20–24) depicted as a continent-spanning conveyor. The ideal is that external as well as internal customers and suppliers should act like successive operations on an assembly line, with neither more nor less than the necessary quantity of work arriving exactly when it is needed. This underscores the need for cooperative and effective customer and supplier relations.

Customer–Supplier Relations

Schragenheim and Dettmer (2001) describes the problem of suboptimization in supply chains, in which supply chain participants' profit motives result in dysfunctional behavior. *Transfer pricing* is meanwhile a frequent cause of conflicts between internal suppliers and customers. A typical supply chain management challenge is to resolve the often-inherent conflict between these two objectives (Holt and Button, 2000, 2):

1. "To maximize the revenue of the entire chain, links must make decisions in the interests of the entire chain."
2. "To protect the interests of the links, the chain must make decisions in the interests of the links."

Linear programming (LP) can at least in theory identify the product or service mixture that will maximize an entire supply chain's profit. It might, however, be necessary to negotiate pricing between supply chain participants to make the optimum solution attractive to everyone (Levinson and Rerick, 2002, 163–166).

The need for porosity—the external relations counterpart to what W. Edwards Deming calls "breaking down organizational barriers"—cannot be overemphasized. As an example, if a customer is dissatisfied with a product or service, can it go directly to the department that produced it or must it go through the sales or marketing department? If a supplied product or service is unsatisfactory, can the internal department that uses it work directly with its supplier or must it work through the purchasing department? If these organizational barriers are present in both the customer and supplier, the complaint must go up the customer's hierarchy to the purchasing department, over to the supplier's sales or marketing department, and down to the producing department. This form of bureaucracy is not conducive to effective supply chain management.

Transportation

Many wars have been won not by having a superior army but rather by having a superior logistics system that could deliver supplies where and when they were needed. Logistics is also a central element of a Lean enterprise that must not be overlooked. Inefficient transportation is a form of waste that runs up the price of the finished good or service.

Even the manner in which trucks, rail cars, and ships are unloaded is an important consideration. A truck or ship that is waiting at a dock to load or unload is not doing anything useful, and it also wastes the driver's or crew's time. The concept that loading and unloading are non-value-adding activities goes back at least to Gilbreth (1911, 55), who recommended, "Two-horse carts with horses changed from the empty to the full carts will require fewer and cheaper motions than any other methods of transportation." The idea is that the motive power, whether it be a team of horses, truck tractor, or railroad engine, need not wait for its cargo to be loaded or unloaded.

Unlike a truck tractor or railroad engine, however, a container ship cannot leave its cargo behind to be unpacked at leisure while it goes off to get another load. Malcom McLean's invention of the intermodal shipping container, in fact, reduced greatly the amount of time ships must spend in port to load and unload. Prior to the invention of the container, "a cargo ship typically would spend as much time in port being loaded and unloaded as it did sailing the seven seas," whereas a steamship executive complained that it cost more to move a cargo 1,000 feet from a pier to a ship's hold than it cost to move the same cargo across the ocean (Cudahy, 2006). It can be argued that the intermodal container is a form of external setup that transfers the non-value-adding work of packing and unpacking a specific cargo from the ship to the container.

Consider also the self-limiting paradigm that a truck must load or unload from the rear. Fruehauf's gull-winged trailer can open on both sides, which allows the removal of an entire load of diesel engines in 15 minutes. The same job would take about two hours with a conventional trailer (Schonberger, 1986, 166). This is the kind of thinking that should go into the logistics system. The same concept applies to services such as passenger transportation. Commuter trains have one or more doors per car, so they need to stop for roughly a minute at each station. Meanwhile, the fact that most passenger planes have customers embark and disembark from a single door wastes the time of the passengers, crew, and capital asset alike.

Schonberger (1986, 162) warns explicitly against the world sourcing fad because of its dysfunctional effects on otherwise-Lean supply chains, a concept on which Walker (2001b) expands as follows: "A supply chain that is completely synchronized downstream within a country or a geographic region will lose its synchronization at some point upstream where the international logistics-driven lot sizing takes effect." Gardner (2001, 29–33) adds,

> For example, in the case of ocean cargo from Asia to the United States, transportation lead-time door-to-door can be in excess of 25 days. In such cases quality must be absolutely reliable. ...once merchandise enters the logistics chain, reversing the process is very difficult and expensive to handle. JIT operations cannot afford to have compromised shipments in the pipeline, particularly when the problems don't surface until they hit the port of destination.

This is simply the logistics chain extension of the basic Lean manufacturing principle that inventory gives poor quality a place to hide until it is discovered at an inopportune time—like when the factory needs the parts to fill an urgent order.

Even if the parts' value-to-weight ratio is high enough to justify shipment by air freight, transportation can still add days to the cycle time. Many semiconductor companies, for example, send chips offshore to Indonesia, Malaysia, Singapore, and Thailand for assembly and packaging because labor is cheaper there. The round trip doubtlessly adds at least two days to the cycle time which can be a competitive disadvantage when customers want their parts *now*.

On a final note, the so-called "JIT warehouse," a place that holds stock until its customer plant needs it, is an oxymoron. Although it keeps the inventory off the suppliers' and customers' premises where no one will notice it, it is nonetheless inventory. As such, it has all the disadvantages inherent in batching, queuing, and providing a hiding place for defects.

Chapter 13

Measurement and Continuous Improvement: Details

Measurement and Analysis for Continuous Improvement

The collection of information for quality and process control purposes is at best a value-assisting activity. Its cost-of-quality category would therefore be Appraisal or Prevention.

Consider, for example, a job in which the torque with which a fastener is tightened is a critical-to-quality characteristic. Most quality auditors would be more than satisfied, and rightly so under straight quality management system (QMS) standards, if the torque gage were calibrated properly and the measurements recorded accurately. Application of the torque gage is, however, almost a repetition of the original job of tightening the fastener, and it is not particularly efficient to do the same job twice.

The obvious solution is a torque wrench that doubles as a gage and records the torque while the worker tightens the fastener. Mountz's TorqueMate® Digital Torque Wrench and Digital Torque Screwdriver have exactly this capability. The general lesson is that a tool that measures the work it does, and preferably while it is doing it, can be extremely useful in eliminating non-value-adding cycle time and/or worker time from the job.

Proactive Action

The need for proactive action cannot be overemphasized because, unlike nonconformities that trigger corrective action as required by ISO 9001:2008 or ISO/TS 16949, an activity that is producing perfect quality may be enormously wasteful. Many forms of waste are asymptomatic because they do not cause the "pain" of rework, scrap, or customer returns.

Standardization and best practice deployment are part of the foundation of Lean manufacturing and Lean enterprise. Frederick Winslow Taylor speculated that ancient trades like bricklaying, carpentry, and so on had been improved many times throughout history. The workers did not, however, record the improved procedures, so the benefits were lost as soon as they retired or died. The only way to ensure continuous improvement is to record the best known way of doing a job and making that method standard for the job. H.L. Gantt (the inventor of the Gantt chart) described standardization as follows: "Standardization consists in reducing to written rules the best methods, and prescribing them for general use" (The System Company, 1911a, 15).

Best practice deployment means sharing improvements with similar activities and operations that might benefit from them. Taylor (1911a, 67) defines very clearly the relationship between continuous improvement, closed loop proactive action, standardization, and best practice deployment:

> It is true that with scientific management the workman is not allowed to use whatever implements and methods he sees fit in the daily practise of his work. Every encouragement, however, should be given him to suggest improvements, both in methods and in implements. And whenever a workman proposes an improvement, it should be the policy of the management to make a careful analysis of the new method, and if necessary conduct a series of experiments to determine accurately the relative merit of the new suggestion and of the old standard. And whenever the new method is found to be markedly superior to the old, it should be adopted as the standard for the whole establishment.

The last sentence embodies both standardization and best practice deployment. The improvement is applied to the job at hand (standardization) and also to similar activities throughout the organization (best practice deployment).

Ford and Crowther (1926, 85) described best practice deployment as follows: "An operation in our plant at Barcelona has to be carried through exactly as in Detroit— the benefit of our experience cannot be thrown away." AIAG's (2006, 137) Lessons Learned database and Read Across/Replicate Process (172–173) prevent loss of the benefits of valuable experience.

Chapter 14

Additional Lean Environmental and Energy Practices

This chapter is intended to be synergistic with the ISO 14001 standard for environmental management systems and ISO 50001 standard for energy management systems. *Compliance with ISO 14001 and 50001 should be not only free but profitable and should easily piggyback onto ISO 9001 or ISO/TS 16949.*

Remember that Henry Ford made enormous amounts of money by either not making environmental waste or by finding uses for waste products. This was during a time when he could have legally thrown into the nearest river whatever wouldn't go up the smokestack. The daily output from Ford's wood distillation plant, which included products like charcoal and wood alcohol, was worth $12,000 a day in 1926. Ford's workers then earned a minimum daily wage of $6 (which was quite high for that time), so distilling the waste wood instead of dumping it in a landfill gave Ford 2,000 free workers. The following statement (Sinclair, 1937, 61) summarizes a key concept of good Lean environmental practices:

> He perfected new processes—the very smoke which had once poured from his chimneys was now made into automobile parts.

Chapter 2 showed that Ford's processes included adsorption and reuse of solvent fumes from a coating operation, and Sinclair may have construed this as the conversion of smoke into automobile parts. The basic concept is, however, the need to pay attention to every waste that might otherwise be taken for granted. Ford (Ford and Crowther, 1926, 124) emphasized this by saying that, although it is indeed impossible to get something from nothing, it is frequently possible to get value from what was considered nothing. This principle should play a central role in a Lean management system (LMS), and in ISO 14001 and ISO 50001.

Identification of Material and Energy Wastes

Material and energy wastes are often asymptomatic like other forms of waste. If the process is making a high-quality product, there is no pain in the form of scrap, rework, or customer returns, and nobody complains. Chapter 2 described how the material and energy balance technique that chemical engineers use to assess processes gives material and energy wastes absolutely no place to hide. "Dumpster diving," or visual observation of everything that is thrown away, is far less rigorous analytically, but it requires almost no training or education whatsoever.

As an example, Ford's observant workers looked at a metal sheet with six holes in it (the product) and asked what became of the metal that was in the holes. The answer was that it was sent back to the blast furnace for recycling, but the workers realized that two of the metal discs would, when pressed together, make a very strong radiator cap. Ford had obviously created and fostered a culture that encouraged and empowered the workforce to identify this kind of waste, and this was among his principal success secrets.

The same concept could easily apply to wastes of energy. If a worker can feel radiated heat from a piece of equipment, it represents energy waste that might (1) be prevented by insulation or (2) somehow captured and put to constructive uses. General Electric recently acquired Calnetix Power Solutions, which uses an organic Rankine cycle (a power generation that uses a low-boiling organic solvent instead of water and steam) to convert relatively low-temperature waste heat into mechanical energy (LaMonica, 2010). The maximum theoretical (Carnot cycle) efficiency for 250°F waste heat is only about 25.3 percent if the cold reservoir is at 70°F, which means the real efficiency of the Rankine cycle will be lower. This is still better than throwing the heat away, though, and the 125-kilowatt figure cited by the reference can add up to real money at 3,000 kilowatt-hours per day.

Another obvious form of waste is the heat that automobiles discharge through their exhaust systems. All the enormous heat that is generated in a catalytic converter currently goes out the tailpipe, but Ford, General Motors, and BMW are researching thermoelectric heat recovery systems that will convert some of this energy into electricity that the vehicle can use for various needs (Ashley, 2011). The lesson here it that if a worker or customer can notice the waste, there is probably a way to mitigate it and put the wasted energy to use.

The next section provides examples of how to reduce material and energy wastes. What is important is not necessarily the individual technique, which may not apply to the reader's industry, but rather the thought process behind it.

Reduction of Material and Energy Wastes

Numerous techniques exist to remove material and energy wastes even from value-adding processes that transform the product. Chapter 2 described how selection of an alloy that did not require heat treatment eliminated not only this cycle time element but also the energy necessary for heat treatment. "Cryogenic Machining Takes Flight with F-35" (2011, 33) describes how cryogenic machining, at least for titanium, not only increases the material removal rate but also extends tool life. It also leaves no spent cutting fluids to become environmental waste management problems; the liquid nitrogen vaporizes harmlessly.

> Cryogenic machining is said to increase cutting-tool life up to a factor of 10 and doubles the material-removal rate, compared to conventional machining methods in certain applications.

Cryogenic machining therefore cuts 50 percent from the cycle time of the indicated operation. Liquid nitrogen is relatively cheap, and it eliminates the ISO 14001 aspect of spent cutting fluids completely. The only conceivable caveats are the safety issues related to the use of a cryogenic liquid, plus its ability to displace oxygen from a confined space. Conventional safety and industrial hygiene practices are more than sufficient to deal with these issues.

This innovative application of liquid nitrogen leads to discussion of supercritical solvents as environmentally friendly alternatives to traditional solvents. The thought process is again more important than the specific applications.

Supercritical Solvents

A supercritical solvent is one whose temperature and pressure both exceed the critical pressure and temperature. The resulting phase is a fluid that cannot be defined either as a liquid or a gas. The critical point itself is alternatively the temperature and pressure at which the liquid and vapor phases become indistinguishable, and the liquid and vapor densities are identical.

Supercritical solvents have very interesting properties and can often perform the same jobs as traditional solvents that would be ISO 14001 environmental aspects. Supercritical carbon dioxide can, for example, decaffeinate coffee and dry-clean clothing, thus taking over jobs normally performed by chlorinated solvents. If the pressure of the solution is allowed to drop, the solute (caffeine or dirt) simply falls out, and the carbon dioxide can then be safely discharged to atmosphere or recovered for reuse.

Counterflow Rinse Systems in Semiconductor Processing and Metal Plating

Semiconductor manufacture uses an enormous amount of deionized water to quench and rinse away other chemicals such as etchants. Peterson et al. (1999) cite a SEMATECH estimate of 1,000 gallons per wafer and with "minimal recycle and reuse across the industry." What is particularly wasteful about these processes is that the waste solution is extremely dilute; it is generally cheaper and easier to deal with a gallon of concentrated waste than two gallons with half the concentration of waste. The reference shows that the chemical concentration model in an overflow rinse tank is that of a continuous stirred tank reactor (CSTR), in which the concentration of the exiting solution is identical to that in the tank. This means each successive gallon of water carries away less of the etchant or the other chemical to be removed.

Chemical engineers prefer the plug flow reactor (PFR), in which reacting fluid elements move through a pipe like miniature batch reactors. The reference says, "Moving the characteristic operating conditions of the rinse tank closer to plug flow conditions can potentially reduce water usage over that required for CSTR conditions," (33) and adds other techniques such as changing the shape of the tank itself. Pulse rinsing meanwhile adds convective removal, i.e., removal due to mechanical flow as opposed to diffusion, of the chemical and also improves efficiency.

Leaching is the dissolution of a solute from an insoluble solid, and extraction is the transfer of a solute from one solvent to another. McCabe and Smith (1976, 607–627) cite the leaching of wax (solute) from waxed paper by kerosene (solvent) and the extraction of acetone (solute) from water (raffinate) by methyl-isobutyl-ketone (MIK, the solvent). The flow of the solvent is generally countercurrent to that of the feed, so the most concentrated feed meets the most concentrated solvent. The idea is to make the exiting solution as concentrated as possible and to therefore use as little solvent as possible. This approach can be roughly imitated by treating the semiconductor wafers as the feed and moving them through successive tanks of water, in which the water flows from one tank to the next in the opposite direction. If there are, for example, two tanks, one gallon does the work of two, so overall water consumption is reduced by half (Figure 14.1).

Counterflow (or cascade) rinsing has been adopted by the semiconductor industry, and it has applications in metal plating operations as well. In metal plating, much of the dragout (plating solution carried from the plating tanks by the parts) can be recovered in a spray rinse tank for recycling to the plating solution tank (Merit Partnership, 1997).

> Using spray nozzles as part of a rinse system can significantly reduce (1) dragout of expensive and hazardous process chemicals and (2) the amount of rinsewater needed. When used on parts over plating and dragout tanks, spray rinses provide a method to recover concentrated process chemicals for reuse. (1)

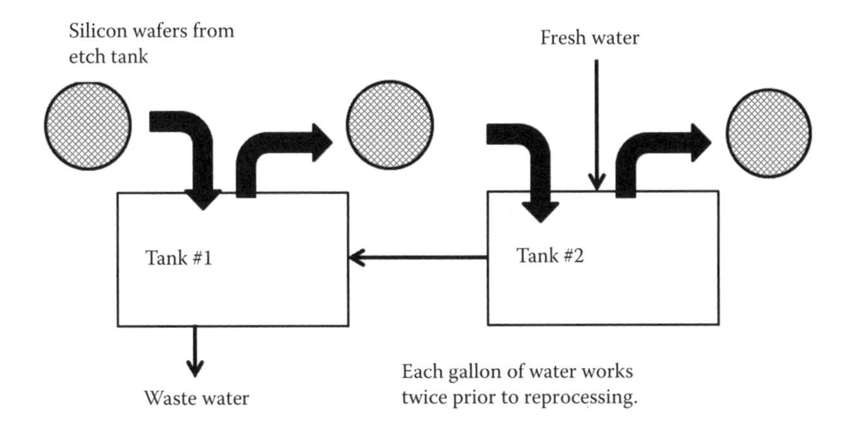

Figure 14.1 Counterflow rinse tanks.

Chapter 3 meanwhile discussed a cadmium electroplating operation (Merit Partnership, 1996). This reference adds that traditional rinsing processes are extremely wasteful and furthermore that dragout wastes the chemical and creates an environmental aspect as well. Redesign of such a process to add a spray rinse to the dragout rinse, both of which return the dragout to the plating tank, followed by a two-stage counterflow rinse tank, reduced water consumption by two-thirds (0.5 versus 1.5 gallons per minute) and cadmium cyanide discharge by half (0.02 ounces per hour versus 0.04 ounces per hour). As a side benefit, the process redesign reduced the number of steps workers had to take from 58 to 21, thus reducing the waste of time of people as well.

The total project cost to improve the cadmium plating operation along with a chromate conversion process was $4,520, whereas the resulting savings in raw materials, water, sewer fees, and wastewater treatment were $2,620/year for a payback period of 1.73 years. Payback is a very conservative engineering economic analysis in contrast to the more rigorous net present value method. The tabulated benefits did not even include the reduction in the amount of walking the workers had to do or the savings in wastewater treatment chemicals. The bottom line is that countercurrent flow systems, in which each volume element of deionized water or rinsewater does two or even more cleaning jobs, are very effective.

Get a Sail!

Chapter 3 described how the *Eugen Maersk* often operates at less than half her top speed to reduce fuel costs enormously. This actually takes us back to the nineteenth century, when speeds of 10 knots or even higher were achievable by sail. The *Preussen* could, for example, achieve a top speed of 20 knots under sail, which was easily competitive with the steam-powered ships of the early twentieth century. Alter (2009) states that the *Preussen's* best average speed between Germany and Chile was 13.7 knots, i.e., faster than the *Eugen Maersk's* 10 knots *while using no fuel whatsoever for propulsion.*

It was common for passersby to yell "Get a horse!" at early motorists whose automobiles had broken down or could not keep pace with equine traffic. The *Preussen* example suggests that "Get a sail!" applies equally to container ships that steam at only 10 knots.

It is not necessary to build or even retrofit container ships with masts for this purpose. Per Rosenberg (2008), *MS Beluga SkySails* deploys a

160-square-meter kite sail that is expected to reduce fuel costs by $1,560 per day. The same article describes the possibility of 5,000-square-meter kites or sky sails, which would presumably be used to propel larger ships. This raises the question as to whether existing container ships could be retrofitted with kite sails, and the German firm SkySails GmbH & Co. KG (www.skysails.info) can doubtlessly provide guidance for anybody who is interested. This company can also take pride in the fact that even Henry Ford, who originated the Green Supply Chain and also the concept of sustainability, did not think of augmenting the engines of his cargo ships with sails.

Don't Ship Air (or Water)

An example of not shipping air involves the use of square instead of circular bottles. One such bottle's label says a square bottle uses less plastic than a circular one of identical volume, which is an error if the walls are of the same thickness. Bottles of equal heights hold equal volumes if their cross-sectional areas are equal, and a circle minimizes the perimeter (and therefore the required material) for any given cross section. The circular container should therefore require *less* plastic than the square container that holds the same volume and is therefore more economical in terms of material.

However, circular containers packed as tightly as possible on a triangular pitch (i.e., with the centers of the circles at the corners of equilateral triangles) occupy about 90.8 percent of the total available area. This can be calculated from the fact that three circles whose centers are on the vertexes of an equilateral triangle have inside the triangle half the area (three 60-degree arcs) of a single circle. Their total area is therefore

$$\frac{1}{2}\pi\frac{D^2}{4} = 0.393D^2$$

The sides of the triangle equal the diameter of a circle, so the triangle's area is

$$\frac{1}{2}\mathit{base} \times \mathit{height} = \frac{1}{2}D \times \left(\frac{D}{2}\tan 60°\right) = 0.433D^2$$

A shipment of square containers can therefore contain 11 percent more product than a shipment of cylindrical containers, which reduces the per-unit energy cost of the shipment. The question is then as to whether the value of the fuel saved through use of square containers exceeds the cost of the extra material necessary for square containers. Another consideration, although it is outside the Lean KPIs, is the fact that the square containers will occupy less retail shelf space than an equal volume of cylindrical containers. A cylinder, on the other hand, offers the most strength, which might not be particularly important for small containers of consumer products but is an obvious safety consideration for drums and tanks of chemicals. The Assess step of IMAIS must consider the economics of tradeoffs of this nature.

Meanwhile, an immediate improvement can be gained through packing circular containers on a triangular pitch instead of a square one (Levinson, 2011b). "Triangular pitch" means the centers of the containers are at the corners of equilateral triangles instead of the corners of squares. Circular containers on a square pitch occupy 78.5 percent of the area in a box (or on a retailer's shelf), whereas, as shown above, those on a triangular pitch occupy 90.8 percent. (This calculation assumes an infinite surface area; in practice, there will be a vacant half-circle in each row for triangular pitch, although the rows can be placed closer together.) Even if square containers are not practical or desirable, an improvement of up to 15 percent is achievable by shipment or storage on a triangular pitch. Incidentally, the same principle applies to the thermoforming of plastic cups from sheets of polymer, depending on the width of the sheets.

The bottom line is that air has no value to anybody other than divers and astronauts, so its shipment on trucks, ships, and railroads should be avoided whenever possible. Note for example that beverages and other liquid products are often packaged where the cans or bottles are made, because shipment of empty cans or bottles is enormously wasteful. Popcorn is far more expensive per pound than unpopped kernels because a shipment of popcorn is primarily a shipment of air. Oatmeal and similar cereals to which hot water is added should be cheaper than puffed cereals for the same reason.

The same principle applies to the shipment of water. Chapter 2 described how Henry Ford tried to avoid the shipment of green wood, or wood with a lot of water content, because it cost money to move the non-value-adding water. Bottled tea is far more expensive than loose or bagged tea because a shipment of bottled tea is essentially a shipment of water.

Innovative Use of Mechanical Energy

This book has provided many examples of how military applications, including the need for motion efficiency and standardization, have driven corresponding applications in civilian enterprises. A friend of Hiram Maxim counseled him during the late nineteenth century: "If you want to make your fortune, invent something to help these fool Europeans kill each other more quickly!"

Recoil had previously been worse than a waste of the firearm's reaction to the bullet's discharge. It was an annoyance to the rifleman and, prior to the invention of the hydropneumatic recoil system at the end of the nineteenth century, it pushed cannons backward and thus required the crews to reposition the weapons after every shot. Only on the wooden warships of previous centuries did recoil serve the useful purpose of moving the guns back to their loading positions. Maxim transformed recoil from a liability into an asset by using it to load and chamber the next cartridge, thus making the first genuine machine guns possible. (Gatling, Nordenfelt, and Gardner guns could fire very rapidly, but they required a soldier to turn a crank or work a lever to operate them.)

This book has already discussed regenerative braking, which recovers mechanical energy as electricity instead of dissipating it as heat. Arnold and Faurote (1915, 312–313) describe how a tool for the placement of radiator tubes into radiator fins put its retrograde motion to good use: "The machine works in both directions so that when the ram is being withdrawn from one set of molds, it is forcing the tubes into the corresponding one on the other side."

Ford and Crowther (1926, 75–76) described how a lathe with two arbors allowed the cutting tool to work on an unfinished piece while the finished one was removed and replaced. The lathe was not taken out of gear (idled) between jobs; instead, the tool had two arbors between which the cutting tool could alternate. This also exemplifies the concept of external setup; the tool need not stop while fresh work is supplied. In addition, the tool's retrograde motion is put to use instead of going to waste.

Economy of Scale in Renewable Energy

The high capital cost of certain forms of renewable energy, especially wind energy, is an entry barrier for homeowners and small business owners. Large wind turbines, however, cost less per kilowatt-hour of capacity than small ones. This suggests a combination of resources, such as purchases of shares, in a big wind turbine from which investors take dividends in kind—specifically credits for electricity—instead of cash. The same concept applies to geothermal energy, in which the cost of a geothermal well can be shared by multiple investors.

Application to Agriculture

Henry Ford, who grew up on a farm, noticed the enormous waste of human labor that went with agriculture:

> I believe that the average farmer puts to a really useful purpose only about 5 per cent of the energy that he spends. If any one ever equipped a factory in the style, say, the average farm is fitted out, the place would be cluttered with men. The worst factory in Europe is hardly as bad as the average farm barn. Power is utilized to the least possible degree. Not only is everything done by hand, but seldom is a thought given to logical arrangement. A farmer doing his chores will walk up and down a rickety ladder a dozen times. He will carry water for years instead of putting in a few lengths of pipe.
>
> ...The moment the farmer considers himself as an industrialist, with a horror of waste either in material or in men, then we are going to have farm products so low-priced that all will have enough to eat, and the profits will be so satisfactory that farming will be considered as among the least hazardous and most profitable of occupations. (Ford and Crowther, 1922, 15–16)

Ford put this into practice on his own farm in Dearborn, Michigan, where tractors performed work that was once done by horses. Even the modern factory farm is, however, enormously wasteful of materials, including water for irrigation and fertilizer. The farmer must ask what portion of the fertilizer he purchases is actually taken up by crops and what portion seeps into the ground or, even worse, becomes an environmental problem due to runoff. What fraction of water that is purchased for irrigation is actually used by the crops, and how much evaporates or is lost in the ground? Even if a detailed material balance is not possible, the concept itself should suggest to the farmer the enormous waste that is present.

In addition, most farms in North America have only one growing season, which translates into waste of the time of things, specifically capital assets for which paying work is available.

The first thought that comes to mind is the hydroponic farm, which can recycle both water and nutrients. The greenhouse environment also protects the crops from droughts, frosts, and similar disasters that can wipe out a harvest, and it can operate year round. The primary drawback appears to be that a catastrophic failure in the hydroponic system itself could wipe out the entire harvest, although this risk can probably be managed with redundant systems, uninterruptable power supplies, and so on. Gotham Greens (Mosher, 2011) has proven that hydroponic farms can be placed on rooftops in urban areas, which (1) creates another source of revenue for the building and (2) reduces or eliminates the cost of transporting food into the city.

However, another concept is the vertical farm, which applies Ford's (Ford and Crowther, 1922, 77) observation about skyscrapers to agriculture:

> A building thirty stories high needs no more ground space than one five stories high. Getting along with the old-style architecture costs the five-story man the income of twenty-five floors.

Ground space or acreage is everything in agriculture, but a 10-story vertical farm that covers one acre can produce as much food as a traditional 10-acre farm—or two times as much, given a controlled environment and year-round operation. Sweco (2011) reports that a vertical greenhouse is being planned for Stockholm, and numerous other examples can be found online. The reference stresses that growing populations and urbanization will require innovative ways to produce food, and that vertical greenhouses are suitable for urban environments.

Hydroponics and vertical farms require far more capital investment than traditional farms, but some of this is offset by elimination of the capital cost of tractors and, in the case of vertical farms, land. Meanwhile, Sweet Water Organics in Milwaukee (Herzog, 2010) has come up with perhaps an even more innovative concept: a largely closed system in which fish (product) provide fertilizer for crops (product), which filter the water in which the fish live and also provide food for worms the fish can eat.

Standard and Davis (1999, 2–3) describe how Adams Citrus Nursery in Haines City, Florida, essentially built an assembly line for citrus trees. Seedlings enter at one end and move very slowly on a rail system—essentially a conveyor belt—to emerge as mature trees nine months later. This is 25 percent of the cycle time for traditional nurseries, which means trees can be grown to order for citrus orchards instead of to forecast.

An idea that comes to mind immediately is acceleration of growth with carbon dioxide–enriched atmospheres and/or sunlamps, which reduce production cycle times even more and also get far more use out of the capital asset (greenhouse or vertical farm). If the rate of a chemical reaction like photosynthesis depends on the concentration of the reactant—in this case, carbon dioxide—a higher concentration increases the rate. The thought process here is extremely important because the chief barrier to widespread implementation seems to be the paradigm that crops are always grown in ambient conditions. This is how humans have grown things since the dawn of agriculture, and only in the past hundred or so years has it become possible to break away from thousands of years of ingrained habits.

ScienceDaily (1998) reports that enrichment of carbon dioxide by 200 ppm, or 56 percent above normal atmospheric levels, increased the yield of cotton plants by 40 percent and the final weight of wheat crops by 20 percent. The website of Home Harvest® Garden Supply says the following about Bibb lettuce: "By adding CO_2 to the atmosphere around the plant, a 40 percent crop increase was achieved. Whereas previous crops averaged 22 heads per basket, lettuce grown in the increased CO_2 atmosphere (550 ppm) averaged 16 heads of better quality per basket."

Innovative Thinking in Transportation

"Say No to Bigger Trucks" (AAAWorld, 2012) opposes a proposed increase on the national weight limit for trucks from 80,000 to 97,000 pounds because (1) more weight increases the braking distance and (2) heavier trucks cause more wear and damage to roads. Fuel consumption does not, however, increase in proportion to the truck's load because air friction remains constant, so bigger loads mean a lower transportation cost per ton of cargo. The cost of the driver's time also is shared by a bigger load, which also reduces the transportation cost.

As with most other situations of this nature, the choice between safer roads and lower transportation costs is not an either/or choice. It is intuitively obvious that the addition of an axle and four tires to the tractor-trailer will (1) distribute the weight over 22 instead of 18 tires, and therefore 22 percent more surface area, and (2) provide another set of brakes. This has in fact been proposed:

> "If we could add an axle and four tires to the trucks delivering finished paper from just one of our mills in the southeast so we could haul 97,000 lbs.—just like in Great Britain—we could reduce the number of weekly truck shipments from 600 down to 450, a reduction of 150 trucks per week," says John Runyan, senior manager of federal government relations for International Paper and co-chair of the Coalition for Transportation Productivity. (Kilcarr, 2009, webpage)

> The Renard train, as our readers are aware, consists of a number of road vehicles—for passengers or goods as the case may be—coupled up behind a locomotive, the power for which, instead of being used only for direct haulage, is transmitted by a shaft right through the train, so that each vehicle, while it can hardly be described as self-propelled, is, at any rate, independently driven. (*Automotor Journal*, 1907, 242)

Automotor Journal adds that the load a tractor (or locomotive) can pull is limited, no matter how powerful the engine might be, by the adhesion or grip of the prime mover's wheels. If, however, the engine can supply power to each car or trailer, the load itself provides the necessary adhesion. This suggests the idea of providing the power electrically instead of mechanically, as was done by the Renard train. The thought process is what is important here, noting that a road train cannot easily negotiate urban or suburban areas no matter how it is powered.

However, another form of road train promises to reduce traffic congestion and save fuel. Schwartz (Fast Company, n.d.) reports that Volvo is developing "platooning" technology that would allow motorists to merge their cars into an electronically controlled convoy lead by a professional driver and allow the leader's computer to take control of their vehicles. This would allow the vehicles to travel safely with very little space between them—it is safe for a computer whose reaction time is measured in milliseconds to tailgate—thus allowing far more vehicles to use the same road without causing congestion.

The ability to decrease the intervals between vehicles without causing a safety problem tells only part of the story. Recall that, in a factory setting, variation in processing and material transfer times increases overall cycle times along with inventory. Variations in vehicle speed, especially when a highway is operating at close to full capacity, can cause traffic slowdowns and stoppages for no apparent reason whatsoever. It is like performing Goldratt's matchsticks-and-dice exercise with cars instead of matchsticks. A computer-controlled convoy or "platoon" has no variation whatsoever in vehicle speed, which suggests that bumper-to-bumper (literally) traffic could flow at normal speeds without any interruption whatsoever.

Schwartz adds that vehicle platooning reduces fuel consumption due to air friction, and also that drivers can use the travel time constructively because they do not have to pay attention to vehicle operation until it is time to leave the convoy. This is the kind of thought process that could ease road congestion, which wastes driver time and vehicle fuel, in places like Los Angeles that are infamous for it.

The next section elaborates on this one with a focus on the 4 Rs: Refuse, Reduce, Reuse, and Recycle. It is primarily applicable to suppression of material waste.

4 Rs: Refuse, Reduce, Reuse, Recycle

These four words are a useful way to remember how to reduce or eliminate material waste.

1. *Refuse* means to refuse to make the waste in the first place, i.e., to prevent it.
2. *Reduce* the waste if it is not possible to prevent its creation.
3. *Reuse* the waste.
4. *Recycle* the waste.

Ford and Crowther (1926, 94) elaborate on refusing to make the waste instead of recycling it: "The ideal is to have nothing to salvage." This book has already described how Design for Manufacture (DFM) can include design of the product to reduce the need for machining and therefore the generation of metal waste and possibly spent cutting fluids. Ford and Crowther (1926, 95) also describe how welding two pieces together to make a cross-shaped oilcan holder instead of stamping the part from sheet metal prevented the generation of recyclable metal waste.

The combustion of coal cannot help but generate sulfur oxides if the coal contains sulfur compounds. Ford and Crowther (1926, 105–106 and 175–176) stated that coal contains valuable chemicals whose use as fuel would be an enormous waste, so Ford converted the coal into coke before he burned it. The coal cost $5 a ton but, after conversion into coke and coal chemicals, was worth $12 a ton, so Ford effectively got his coal for less than nothing. The sulfur, after conversion into ammonium sulfate, could be sold as fertilizer and did not become what is recognized today as an environmental problem. It would not have even been necessary (under modern environmental protection laws) to scrub the sulfur dioxide out of the stack gases, because it would not have been there in the first place.

The decision to discard or reuse/recycle is, of course, a function of the relative cost of paying for disposal versus the benefits of recycling. Consider the ostensibly virtuous goal of recycling polystyrene foam peanuts. The cost of transporting them (noting that they are mostly air) to a recycling center may exceed anything the center might pay for them, if it pays at all. If the means of transportation involves an internal combustion engine, it generates pollution and, for those who consider it a problem, carbon dioxide in the bargain. It may, in fact, be more cost effective to pay whatever a garbage collector charges to dump them in a landfill.

Foam peanuts and similar packaging materials are, however, used so frequently that it makes no sense to discard them if it can be avoided. The most obvious way to reuse them is for the company that receives them to use them to package its own products. If this is not feasible, it might make sense for businesses in an industrial park to post availability and needs for packaging materials. This would eliminate the disposal cost for whoever wanted to get rid of them, and the price to the company that needed them would be that of sending a truck over to get them. The Environmental Information Exchange (Oxford Brookes University) offers a "waste exchange" in the United Kingdom for this express purpose, and it includes wastes other than packaging materials.*

The same reference reports that the ubiquitous wooden pallets that are common in the shipping and warehousing industry can be reused or recycled instead of being thrown away:

---

* http://www.brookes.ac.uk/other/eie_old/wastex.htm

> Most companies will take damaged pallets that are repairable, and some will collect pallets that are beyond repair as part of a load of better condition ones or for a fee. These—and non-standard ones—can still be used as "spare parts" for repairs, or they can be chipped for fuel, chipboard or for the horticultural market.*

The reference adds, however, that painted or preservative-treated pallets are less easy to recycle at the end of their useful lives as pallets.

---

* http://www.brookes.ac.uk/other/eie_old/recpals.htm

Chapter 15

Conclusion

This book's purpose has been to provide the user with a framework for a sustained and disciplined approach to continuous Lean improvement. The key takeaways are as follows:

1. A set of simple, straightforward, and easy-to-remember *Lean key performance indicators* (KPIs) that are also the *critical to Lean* (CTL) characteristics of any process that delivers a product or a service
2. The Isolate, Measure, Assess, Improve, Standardize (IMAIS) continuous improvement cycle that, although similar to all other improvement cycles, sets out explicitly to look for trouble in the form of waste instead of waiting for trouble to be brought to it by, for example, a corrective action request
3. An unofficial and therefore customizable standard that deploys the Lean KPIs with focus on the same business processes that ISO 9001 addresses

The next sections will summarize each of these elements in more detail.

Lean KPIs and CTL Characteristics

Remember that ISO 9001:2008 and ISO/TS 16949 focus primarily on critical to quality (CTQ) characteristics, and that these are almost universally characteristics of the product or service as opposed to the process that delivers them. CTL characteristics are in contrast inherent to the process itself, and they consist solely of efficiencies in terms of (1) time, (2) materials, and (3)

Table 15.1 Critical to Quality versus Critical to Lean

	Process CTL Characteristics	
Product CTQ Characteristics	*Lean*	*Wasteful*
Good (conforming)	Inexpensive quality output	Expensive quality output
Poor (nonconforming)	Cheap junk	Expensive junk

energy. Table 15.1 shows four possible situations in terms of CTQ and CTL characteristics.

Capable processes with good controls will deliver output whose CTQ characteristics meet customer requirements, and the focus of the quality management system (QMS) is to ensure that the processes (including not only production but supporting activities like purchasing, measurements, and documentation) will do this. However, whether this high-quality output is inexpensive or costly depends on the CTL characteristics of the process. The output of a quality-deficient process won't meet the customer's requirements and is largely useless whether it is cheap or expensive. This fact reiterates the need to build the Lean management system (LMS) on a solid QMS.

Chapter 2 showed that "Lean" and "wasteful" are almost entirely quantifiable by exactly four KPIs or CTL characteristics: (1) waste of the time of things, (2) waste of the time of people, (3) waste of materials, and (4) waste of energy. Only when all these wastes are zero is it possible to conclude that there is no better way to do the job in question, which means it is always a good idea to assume that there is a better way. The next step is to put these measurements to work to identify and remove waste.

IMAIS Improvement Cycle

IMAIS deploys the Lean KPIs by (1) isolating the process for analytical purposes with a control surface or control envelope, and (2) measuring all inputs and outputs of time, material, and energy. Cycle time accounting can assess very rigorously the time the work spends in the process while it distinguishes value-adding from non-value-adding time. No waste of material or energy can hide from the equally if not more rigorous material and energy balance technique. Traditional motion efficiency and human factors methods can identify waste of the time of people.

Once the wastes are visible, traditional root cause analysis and experimentation can identify the reasons for the wastes and possible ways to remove them. Implementation and verification of the improvements followed by standardization and best practice deployment completes the improvement cycle. The next step is to apply this approach through all relevant business processes.

LMS:2012

The unofficial standard is designed to piggyback onto ISO 9001:2008 and to address the same business, manufacturing, and service processes that the quality standard addresses. These include specifically:

1. The management system, in this case as it applies to the Lean KPIs
2. Organizational responsibility, including reciprocal responsibilities of management, the workforce, and supply chain partners
3. Resources, including human and physical ones
4. The realization and delivery of the product or service
5. Measurement, auditing, and continuous improvement

Diligent adherence to the unofficial standard ensures continuity of the LMS, including efforts to continuously remove all forms of waste from every process. This prevents backsliding or, in the event of a catastrophic turnover of leadership personnel, outright loss of the Lean culture and everything that goes with it.

Bibliography

AAA World. 2012. "Say No to Bigger Trucks." January/February 9.

Alter, Lloyd. 2009. "Slow Freight: Sail Power Is Actually Faster than Containerships Today." April 13. http://www.treehugger.com/renewable-energy/slow-freight-sail-power-is-actually-faster-than-containerships-today.html (accessed April 9, 2012).

Arnold, Horace Lucien, and Fay Leone Faurote. 1915. *Ford Methods and the Ford Shops*. New York: *The Engineering Magazine*. Reprinted 1998, North Stratford, NH: Ayer Company Publishers, Inc.

Ashley, Steven. 2011. "Latest Out of Tailpipes: Electricity with Exhaust." *New York Times*, July 29. http://www.nytimes.com/2011/07/31/automobiles/latest-out-of-tailpipes-electricity-with-exhaust.html (accessed April 9, 2012).

Automotive Industry Action Group (AIAG). 2006. "CQI-10: Effective Problem Solving: A Guideline." Available from aiag.org.

Automotor Journal. 1907. "The Renard Road Train." February 23, 242–243. http://books.google.com/books?id=fm4nAAAAMAAJ&lpg=PA243&ots=jPSyNNkHmS&dq=1907%20%22the%20auto%22%20road%20train&pg=PA242#v=onepage&q=1907%20%22the%20auto%22%20road%20tr-ain&f=false (accessed April 9, 2012).

Bakker, Robert M. 1996. "Why Companies Fail Quality Audits." *Manufacturing Engineering*, May http://www.highbeam.com/doc/1P3-9717939.html (accessed April 9, 2012).

BBC News, 2009. "China Unveils Emissions Targets Ahead of Copenhagen." November 26. http://news.bbc.co.uk/2/hi/8380106.stm (accessed April 9, 2012).

Bennett, Harry, as told to Paul Marcus. 1951. *Ford: We Never Called Him Henry*. New York: Tom Doherty Associates, Inc.

Blanchard, David. 2007. "Census of U.S. Manufacturers: Lean Green and Low Cost." *Industry Week*, October. http://www.industryweek.com/articles/census_of_u-s-_manufacturers_--_lean_green_and_low_cost_15009.aspx (accessed April 9, 2012).

Block, Marilyn R., and I. Robert Marash. 1999. *Integrating ISO 14001 into a Quality Management System*. Milwaukee, WI: ASQ Quality Press.

Brassey's Naval Annual. 1899. http://www.gwpda.org/naval/brassey/b1899o06.htm (accessed April 9, 2012).

Bryant, Nick. 2004. "Maldives: Paradise Soon to be Lost." BBC News, July 28. http://news.bbc.co.uk/2/hi/south_asia/3930765.stm (accessed April 9, 2012).

Byron, George Gordon. 1857. *Don Juan*. Halifax: Milner and Sowerby. http://www.gutenberg.org/files/21700/21700-h/21700-h.htm (accessed April 9, 2012).

"California's Biofuel Rules Rejected by Judge." *San Francisco Chronicle*. 2011. December 30. http://www.sfgate.com/cgi-bin/article.cgi?f=/c/a/2011/12/29/BUBR1MIEOI.DTL (accessed April 9, 2012).

Caravaggio, Michael. 1998. "Total Productive Maintenance." In *Leading the Way to Competitive Excellence: The Harris Mountaintop Case Study*, edited by William A. Levinson. Milwaukee, WI: ASQ Quality Press.

Cianfrani, Charles, and Jack West. "2010. ISO 9001:2008 Explained." American Society for Quality, recorded Webinar (accessed June 5, 2010).

Clausewitz, Carl von. 1976. *On War*. Translated by M. Howard and P. Paret. Princeton, NJ: Princeton University Press.

"Cryogenic Machining Takes Flight with F-35." *Manufacturing Engineering*, 2011. November, 33: 35 (accessed April 9, 2012).

Cubberly, William H., and Ramon Bakerjian, eds. 1989. *Tool and Manufacturing Engineers Handbook, Desk Edition*. Dearborn, MI: Society of Manufacturing Engineers.

Cudahy, Brian J. 2006. "The Containership Revolution: Malcom McLean's 1956 Innovation Goes Global." *TR News*, September–October 2006, 5–9. http://onlinepubs.trb.org/onlinepubs/trnews/trnews246.pdf

Dobb, F. P. 2004. *ISO 9001:2008 Quality Registration Step by Step*. London: Elsevier Butterworth Heinemann.

Emerson, Harrington. 1912. *Efficiency as a Basis for Operation and Wages* 3rd. edition. New York: The Engineering Magazine. http://books.google.com/books?id=ARg9AAAAYAAJ&printsec=frontcover&dq=%22Efficiency+as+a+Basis+for+Operation+and+Wages%22+emerson&hl=en&sa=X&ei=tbyET8XEMurw0gHY3N3xBw&ved=0CDMQuwUwAA#v=onepage&q=%22Efficiency%20as%20a%20Basis%20for%20Operation%20and%20Wages%22%20emerson&f=false (accessed April 26, 2012).

Environmental Information Exchange. Oxford Brookes University. http://www.brookes.ac.uk/other/eie_old/wastex.htm (accessed March 7, 2012).

Environmental Information Exchange. Oxford Brookes University. http://www.brookes.ac.uk/other/eie_old/recpals.htm (accessed April 9, 2012).

Ford, Henry, and Samuel Crowther. 1922. *My Life and Work*. New York: Doubleday, Page & Company.

Ford, Henry, and Samuel Crowther. 1926. *Today and Tomorrow*. New York: Doubleday, Page & Company. (Reprint available from Productivity Press, 1988.)

Ford, Henry, and Samuel Crowther. 1930. *Moving Forward*. New York: Doubleday, Doran, & Company.

Ford News. 1922. Vol. 3, 1. Digitized by Google Books.

Ford News. 1922–1923. Digitized by Google Books.

Fraser, John, and Khurshed Kutky (QMI-SAI Global). 2011. "Integrating OSHAS 18001 and ISO 14000 with ISO 9000." June 6. American Society for Quality Webinar.

Gardner, Daniel. 2001. "Movers and Shapers: The Impact of Logistics on Global Supply Chains." *APICS— The Performance Advantage*. May, 29–33.

Home Harvest Garden Supply. (n.d.) http://homeharvest.com/carbondioxideenrichment.htm (accessed April 30, 2012).

Gardner, Les, and Frank Nappi. 2011. "The Total Impact of Minor Stoppages." The 6th Annual Lean Management and TPM Conference, sponsored by Productivity Inc. October 25–26. Dearborn, MI.

George, Michael. 2002. *Lean Six Sigma: Combining Six Sigma Quality with Lean Production Speed*. New York: McGraw-Hill.

Gilbreth, Frank. 1911. *Motion Study*. New York: D. Van Nostrand Reinhold.

Gillibrand, Kirsten. 2009. "Cap and Trade Could be a Boon to New York." *Wall Street Journal*, October 21. http://online.wsj.com/article/SB10001424052748704500604574481812686144826.html (accessed April 9, 2012).

Gilligan, Andrew. 2009. "Copenhagen Climate Summit: 1,200 Limos, 140 Private Planes and Caviar Wedges." *The Telegraph*, December 5. http://www.telegraph.co.uk/earth/copenhagen-climate-change-confe/6736517/Copenhagen-climate-summit-1200-limos-140-private-planes-and-caviar-wedges.html (accessed April 9, 2012).

Glader, Paul. 2006. "It's Not Easy Being Lean." *Wall Street Journal*, June 19: B1.

Goldratt, Eliyahu, and Jeff Cox. 1992. *The Goal*. Croton-on-Hudson, NY: North River Press.

Good Morning America. 2006. "Corporations Send Employees to Pit Crew U." ABC. April 9. http://abcnews.go.com/GMA/Business/story?id=1817896

Halpin, J. F. 1966. *Zero Defects*. New York: McGraw-Hill.

Harrington, James. 2005. "Lost in the Quality Service Void." *Quality Digest,* November 2005. http://www.qualitydigest.com/nov05/columnists/jharrington.shtml (accessed April 9, 2012).

Harry, Mikel, and Richard Schroeder. 2000. *Six Sigma: The Breakthrough Management Strategy Revolutionizing the World's Top Corporations*. New York: Currency.

Herzog, Karen. 2010. "Perch Return to Local Waters—In an Old Factory." *Milwaukee-Wisconsin Journal Sentinel,* February 5. http://www.jsonline.com/news/milwaukee/83610787.html (accessed April 9, 2012).

Himmelblau, David M. 1967. *Basic Principles and Calculations in Chemical Engineering*. Englewood Cliffs, NJ: Prentice Hall.

Hogg, Dave. 2011. "Lean in a Changed World." *Manufacturing Engineering*, September, 102–113.

Holt, James R., and Scott D. Button. 2000. "Sharing the Destiny Across Multiple Business Units: The Supply Chain Solution." *SIG Synergy*, APICS, September.

Hopp, Wallace J., and Mark L. Spearman. 2000. *Factory Physics,* 2nd ed. New York: McGraw-Hill.

Hoyer, R. W. 2001. "Why Quality Gets an 'F': What Happens When the Bottom Line Overrides a Focus on Customer Needs." *Quality Progress*, October, 32–36.

Imai, Masaaki. 1997. *Gemba Kaizen*. New York: McGraw-Hill.

Inman, Mason. 2006. "Legendary Swords' Sharpness, Strength From Nanotubes, Study Says." *National Geographic News*, November 16. http://news.nationalgeographic.com/news/2006/11/061116-nanotech-swords.html (accessed April 9, 2012).

James, Peter, and Nick Thorpe. 1994. *Ancient Inventions*. New York: Ballantine Books.

Juran, Joseph M. and Gryna, Frank. 1988. *Juran's Quality Control Handbook*, 4th ed. New York: McGraw-Hill.

Kalpakjian, Serope. 1985. *Manufacturing Processes for Engineering Materials*. Reading, MA: Addison-Wesley.

Keegan, John. 1987. *The Mask of Command*. New York: Penguin Books.

Kilcarr, Sean. 2009. "Lifting the Load: Truck Weight Limits." *FleetOwner*, April 1. http://fleetowner.com/management/feature/truck-weight-limit-debate-0409/index.html (accessed April 9, 2012).

Kipling, Rudyard. 1897. *Captains Courageous*. http://www.gutenberg.org/files/2186/2186-h/2186-h.htm

LaMonica, Martin. 2010. "GE Buys Company That Turns Waste Heat into Power." Cnet.com, October 4. http://news.cnet.com/8301-11128_3-20018419-54.html (accessed April 9, 2012).

Ledwith, Paul A. 2004. "Recent Advances in Spin Coating Technology." University of Maryland. Department of Materials Science and Engineering. http://www.mse.umd.edu/undergrad/465_microprocessing_of_materials/465_spring_2004/465_spring_2004_final_project_results/Ledwith.spin%20coating.enma465spr04.pdf (accessed April 9, 2012).

Levinson, William. 2002a. *Henry Ford's Lean Vision: Enduring Principles from the First Ford Motor Plants*. Portland, OR: Productivity Press.

Levinson, William. 2002b. "The Long and Shoremen of It." *Capitalism Magazine*, October. http://www.ideasinactiontv.com/tcs_daily/2002/10/the-long-and-shoremen-of-it.html (accessed April 9, 2012).

Levinson, William. 2005. "Waste Management: Using a Bill of Outputs to Eliminate Excess." *APICS, The Performance Advantage*, January: 33–35.

Levinson, William. 2007. *Beyond the Theory of Constraints*. New York: Taylor and Francis.

Levinson, William. 2009. "A Bailout Can't Fix the Auto Industry's Basic Problems." *Manufacturing Engineering*, June. http://www.sme.org/Tertiary.aspx?id=21450&terms=%22A%20Bailout%20Can%27t%20Fix%20the%20Auto%20Industry%27s%20Basic%20Problems.%22, (accessed April 9, 2012).

Levinson, William. 2011a. "Six Sigma Soup." *Quality Digest* online, January. http://www.qualitydigest.com/inside/quality-insider-column/six-sigma-soup.html.

Levinson, William. 2011b. "Don't Ship or Stock Air." *Quality Digest* online, November. http://www.qualitydigest.com/inside/quality-insider-column/don-t-ship-or-stock-air.html (accessed April 9, 2012).

Levinson, William, and Raymond Rerick. 2002. *Lean Enterprise: A Synergistic Approach to Minimizing Waste*. Milwaukee, WI: ASQ Quality Press.

Levinson, William, and Frank Tumbelty. 1997. *SPC Essentials and Productivity Improvement: A Manufacturing Approach*. Milwaukee, WI: ASQ Quality Press.

Long, Jeff. 2003. "Plant Closing Hits 8,600 Jobs." *Chicago Tribune*, January 15. http://articles.chicagotribune.com/2003-01-15/news/0301150144_1_job-training-motorola-harvard

Lorenzen, Jerry. 1992. "Quality Function Deployment." American Society for Quality Mid-Hudson Section meeting, May 26.

LRQA Business Assurance, 2011. "An Interview with Edwin Pinero—ISO 50001 Independent Chair. Energy Efficiency, the Forgotten Fuel Source, and the Importance of ISO 50001." http://www.lrqa.com/Images/ED%20PINERO%20INTERVIEW%20ISO%2050001_tcm152-222759.pdf

McCabe, Warren, and Julian C. Smith. 1976. *Unit Operations of Chemical Engineering*. New York: McGraw-Hill.

Mege, Claude Jean. 2000. "Is There a HAM in Your Future?" *Manufacturing Engineering*, July: 114–124.

Merit Partnership Pollution Prevention Project for Metal Finishers. 1996. "Modifying Tank Layouts to Improve Process Efficiency." http://www.epa.gov/region9/waste/p2/projects/metal-tanklay.pdf (accessed April 9, 2012).

Merit Partnership Pollution Prevention Project for Metal Finishers. 1997. "Reducing Dragout with Spray Rinses." http://www.nmfrc.org/pdfs2/metal-spray.pdf

Miller, James P. 2002. "Maytag to Shutter Plant in Galesburg." *Chicago Tribune*, October 12. http://articles.chicagotribune.com/2002-10-12/business/0210120209_1_maytag-shares-plant-appliance (accessed April 9, 2012).

Miller, John W. 2009. "Shippers Taking it Slow in Bad Times." *Wall Street Journal*, April 8. http://online.wsj.com/article/SB123913890018398337.html (accessed April 30, 2012).

Mosher, Dave. 2011. "High-Tech Hydroponic Farm Transforms Abandoned Bowling Alley." Wired.com, October 27. http://www.wired.com/wiredscience/2011/10/gotham-greens-hydroponic-farm/ (accessed April 9, 2012).

Munro, Roderick. 2004. *Automotive Internal Auditor Pocket Guide: Process Auditing to ISO/TS 16949:2002*. Milwaukee, WI: ASQ Quality Press.

Norwood, Edwin P. 1931. *Ford: Men and Methods*. Garden City, NY: Doubleday, Doran & Company, Inc.

O'Grady, Mary Anastasia. 2007. Tortilla Facts. *Wall Street Journal*, January 29: A16.

O'Keefe, William. 2009. "Next Bernie Madoff? Emissions Cap-and-Trade Aids the Corrupt, Hurts the Little Guy." *U.S. News & World Report*, April 13 http://www.usnews.com/opinion/articles/2009/04/13/next-bernie-madoff-emissions-cap-and-trade-aids-the-corrupt-hurts-the-little-guy (accessed April 10, 2012).

Ohno, Taiichi. 1988. *Toyota Production System: Beyond Large-Scale Production*. Portland, OR: Productivity Press.

Peters, Tom. 1987. *Thriving on Chaos*. New York: Harper and Row.

Peterson, Thomas W., and Andy Hebda (University of Arizona), Roche, Thomas (Motorola), and Hansen, Eric (Santa Clara Plastics) 1999. "DI Water Reduction in Rinse Processes." University of Arizona.

PRWeb. 2011. "Study Shows 89% Infection Reduction at Miami Children's Hospital." April 4. http://www.prweb.com/releases/2011/4/prweb8266458.htm (accessed April 10, 2012).

Quality Digest. 2011. "International Organization for Standardization Publishes Six Sigma Methodology As Two-Part Standard." September 19.

Rattner, Steven. 2011. "The Great Corn Con." *New York Times*, June 24. http://www.nytimes.com/2011/06/25/opinion/25Rattner.html (accessed April 9, 2012).

"Reflection on Metal Fatigue." *Metal Finishing News*. 2009. July. http://www.mfn.li/article/?id=719 (accessed April 9, 2012).

Resnick, L. 1920. "How Henry Ford Saves Men and Money." *National Safety News*, September 13.

Robinson, Alan, ed. 1990. *Modern Approaches to Manufacturing Improvement: The Shingo System*. Portland, OR: Productivity Press.

Rosenberg, Steve. 2008. "Gone with the Wind on 'Kite Ship.'" *BBC News*, January 23. http://news.bbc.co.uk/2/hi/europe/7205217.stm (accessed April 9, 2012).

Sapsford, Jathon. 2005. "Shifting Gears: As Japan's Elderly Ranks Swell, Toyota Sees New Path to Growth." *Wall Street Journal*, December 21: A1.

Schonberger, Richard J. 1986. *World Class Manufacturing*. New York: The Free Press.

Schragenheim, Eli, and H. William Dettmer. 2001. "Constraints & JIT: Not Necessarily Cutthroat Enemies." *APICS— The Performance Advantage*, April: 57–60.

Schwartz, Ariel (n.d.). "Road Trains—Not Driverless Cars—Are the Future of Hands-Free Driving." Co. Exist, published by Fast Company. http://www.fastcoexist.com/1678130/road-trains-not-driverless-cars-are-the-future-of-hands-free-driving (accessed April 9, 2012).

ScienceDaily. 1998. "High on Carbon Dioxide, Crops of Tomorrow May Yield More Grain." July 15. http://www.sciencedaily.com/releases/1998/07/980715084612.htm.

Shingo, Shigeo. 1985. *Zero Quality Control: Source Inspection and the Poka-Yoke System*. Translated by Andrew P. Dillon. Portland, OR: Productivity Press.

Shingo, Shigeo. 1987. *The Sayings of Shigeo Shingo: Key Strategies for Plant Improvement*. Translated by Andrew Dillon. Portland, OR: Productivity Press.

Shingo, Shigeo. 2009. *Fundamental Principles of Lean Manufacturing*. Bellingham, WA: Enna Products Corporation.

Shirouzu, Norihiko. 2001. "Job One: Ford Has Big Problem Beyond Tire Mess: Making Quality Cars." *Wall Street Journal*, May 25: A1, A6.

Shreve, R. Norris, and Joseph A. Brink. 1977. *Chemical Process Industries*, 4th ed. New York: McGraw-Hill.

Sinclair, Upton. 1937. *The Flivver King*. Second printing, 1987. Chicago: Charles H. Kerr Publishing Company.

Smith, Jim. L. 2011a. "Document Control Is a Cornerstone of Effective QMS." *Quality Magazine*, August: 40–43.

Smith, Jim L. 2011b. "Quality Professionals Must be Multilingual." *Quality Magazine,* December: 12.

Smith, Wayne. 1998. *Time Out: Using Visible Pull Systems to Drive Process Improvements*. New York: John Wiley & Sons.

Society of Automotive Engineers. 1999. Surface Vehicle Recommended Practice J4001, "Implementation of Lean Operation User Manual."

Standard, Charles, and Dale Davis. 1999. *Running Today's Factory: A Proven Strategy for Lean Manufacturing*. Cincinnati: Hanser Gardner Publications.

Strassel, 2007. "If the Cap Fits." *Wall Street Journal,* January 26: A10.

Stuelpnagel, T. R. 1993. "Déjà Vu: TQM Returns to Detroit and Elsewhere." *Quality Progress*. September: 91–95.

Subaru (2011). October. http://www.subaru.com/company/environment-sustainability.html (accessed April 9, 2012).

Sweco. 2011. "Vertical Greenhouse Could be a Reality in Stockholm." http://www.swecogroup.com/en/sweco-group/Press/News/2011/Vertical-greenhouses-could-be-a-reality-in-Stockholm/ (accessed April 9, 2012).

Tadsen, Brent. 2008. "Planning and Executing a Sustainable Lean Transformation." *Industry Week* Webinar, July 17. (W. Levinson's notes made during presentation.)

Taylor, Frederick Winslow. 1911a. *The Principles of Scientific Management*. New York: Harper Brothers. 1998 republication by Dover Publications, Inc., Mineola, NY.

Taylor, Frederick Winslow. 1911b. *Shop Management*. New York: Harper & Brothers Publishers.

The System Company. 1911a. *How Scientific Management Is Applied*. London: A. W. Shaw Company, Ltd.

The System Company. 1911b. *How to Get More Out of Your Factory*. London: A. W. Shaw Company, Ltd.

Von Steuben, Friedrich Wilhelm. 1779. "Regulations for the Order and Discipline of the Troops of the United States." *Manual of Arms*. http://www.2nc.org/steub-man.htm (accessed April 10, 2012).

Walker, Bill. 2001a. "Supply Chain Management." APICS meeting, Pittston, PA, March 14.

Walker, William. 2001b. "Synchronized for Growth." *APICS— The Performance Advantage*, April: 26–29.

Weeks, J. Bruce. 2011. "Is Six Sigma Dead? If It Is, How Can We Revive It?" *Quality Progress,* October: 22–28.

Womack, James P., and Daniel T. Jones. 1996. *Lean Thinking*. New York: Simon & Schuster.

Index

Page references in **bold** refer to tables.

F

G

J

K

L